BestMasters

Springer awards „BestMasters" to the best master's theses which have been completed at renowned universities in Germany, Austria, and Switzerland.
The studies received highest marks and were recommended for publication by supervisors. They address current issues from various fields of research in natural sciences, psychology, technology, and economics.
The series addresses practitioners as well as scientists and, in particular, offers guidance for early stage researchers.

Leonid Ryvkin

Observables and Symmetries of n-Plectic Manifolds

Springer Spektrum

Leonid Ryvkin
Bochum, Germany

BestMasters
ISBN 978-3-658-12389-5 ISBN 978-3-658-12390-1 (eBook)
DOI 10.1007/978-3-658-12390-1

Library of Congress Control Number: 2015960848

Springer Spektrum

Printed on acid-free paper

Springer Spektrum is a brand of Springer Fachmedien Wiesbaden
Springer Fachmedien Wiesbaden is part of Springer Science+Business Media
(www.springer.com)

Acknowledgements

I would like to express my deepest appreciation to my advisor Professor Dr. Tilmann Wurzbacher, who has been an ideal advisor for various reasons. Not only has he proposed the highly interesting topic of the thesis and given me optimal preparation to attacking it through the courses he offered during the last few years, but he also invested a tremendous amount of time in guiding and supporting me in the course of the composition of this thesis. Without his patience, mathematical (and didactic) skill I would have never been able to master this thesis. I would also like to thank him for his enthusiasm and sense of humour which provided "a light in the dark" during the more technical parts of work.

I would like to express my gratefulness to the Gerhard C. Starck Stiftung for the financial and moral support during my studies. Thanks for giving me the possibility to fully concentrate on the subject, that I am so passionate about, and for the exceptional people that I had the privilege to meet through the Stiftung.

I would also like to thank my family and Leonie, who have been a great emotional and practical support during the course of the thesis, who had to endure a lot of vacuum, where I should have been in the last few months. Their care and sympathy were of great value for me.

Last but not least I would like to thank my friends, who have turned the past 4 years of studies into an unforgettable time. It is thanks to them, that studying was fun from the first day to the last. Special thanks goes to Stefan Heuver, who has shown great understanding and kindness during the last few months. Thanks for his great advices, as for instance "If you worry, just stop worrying". Thanks for showing me that my problems were not all that bad, by explaining his problems and last but not least, thanks for forcing me to have a little break once in a while.

Leonid Ryvkin

Contents

1 Introduction

The aim of this thesis is the analysis and understanding of n-plectic (also called multisymplectic) manifolds, their observables and their symmetries.

The development of a geometric theory of n-plectic manifolds naturally starts with the classical case of symplectic manifolds. In that setting one considers a manifold M equipped with a non-degenerate closed two-form ω. This two-form relates $C^\infty(M)$, the space of smooth functions on M with the vector fields on M. This relation induces a Lie algebra structure on $C^\infty(M)$, which, endowed with this structure, is called the Lie algebra of observables of M. The infinitesimal version ζ of a symplectic group action on (M, ω) can often be linearly lifted to this Lie algebra. Such group actions are called weakly Hamiltonian and are the basis of what might be called "momentum geometry", which is also of great interest for physics. The existence of an equivariant moment map is obstructed by a certain cohomology class.

In the general n-plectic case we consider a manifold M with a closed, non-degenerate $n+1$-form. Unlike the symplectic (1-plectic) case, the non-degeneracy of ω only guarantees injectiveness of the map $v \mapsto \iota_v\omega$ but no surjectiveness. Also the space of observables carries a more general algebraic structure which turns it into a "Lie n-algebra", which is a Lie algebra only when $n = 1$.

Lie n-algebras are a special case of L_∞-algebras, whose study constitutes the first chapter of this thesis. Lie n-algebras from n-plectic manifolds are then introduced in Chapter two. The generalization of the notion of a (co-)moment map is the main theme of the third chapter, culminating in an explicit calculation of the obstruction classes to strongly Hamiltonian n-plectic group actions.

The thesis is structured in the following way:

- Sections 1.1 and 1.2 develop two possible perspectives on L_∞-algebras. The earlier one describes an L_∞-algebra as a $\mathbb{Z}$-graded vector space, with a family of brackets $\{l_k | 1 \leq k\}$ satisfying an infinite series of multi-bracket identites. The latter standpoint views an L_∞-algebra as a graded vector space V endowed with a degree one coderivation $D : S^\bullet(sV) \to S^\bullet(sV)$, which squares to zero, where $S^\bullet(sV)$ denotes the cofree graded commutative coalgebra of the vector space sV, where sV is V with degrees shifted by 1. The main result of these sections, the equivalence of these descriptions (Theorem 2.14) is based on the proof sketched in [7], but we work out the full details of the proof here.
- Section 1.3 then discusses the right notion of morphisms of L_∞-algebras. Though L_∞-morphisms more tangible from the coalgebra standpoint, the multi-bracket formulation is more practical for calculations. Hence, the main result of this section is the multi-bracket formulation of L_∞-morphisms (Lemma 2.18). It is based on the calculations made in Appendix A of [5] and results in a formula already stated in in [1], but without proof. We giva a detailed proof and we are also more explicit about the precise range of the summations involved.

- Section 1.4 gives an introduction to the representation theory of L_∞-algebras according to [6]. It presents two perspectives on representations: via L_∞-morphisms into the endomorphism space of a differential graded vector space and via L_∞-modules. The equivalence of both approaches is only quoted from [6] in this thesis.

- Section 1.5 then reduces the "heavy machinery" developed in the preceeding sections to the case of gounded Lie m-algebras, which are of special interest as they include the L_∞-algebras formed by the observables of a multisymplectic manifold. This section mostly reproduces results from [5], where the property "to be grounded" is referred to as "having Property (P)".

- Section 2.1 introduces the notion of n-plectic (or multisymplectic) manifolds and derives their basic properties in accordance with [11]. The discussion of the pre-n-plectic case mostly relies on [5]. The Lie n-algebra of observables is described for both cases.

- Sections 3.1 and 3.2 develop the theory of Hamiltonian group actions on n-plectic manifolds, guided by the classical case described e.g. in [9]. The (homotopy) co-moment map which is the basic object of our discussions was first introduced in [5]. The main results of these sections are Lemma 4.7, which describes the obstruction for an n-plectic action to be weakly Hamiltonian and Theorem 4.13, the which gives a cohomological description of the obstruction for a weakly Hamiltonian action to be strongly Hamiltonian. The latter is a refinement of Theorem 9.7 in [5]. Furthermore we show that the first resp. the last component of a strong homotopy co-moment map yields a covariant multimoment map in the sense of [3] resp. a multi-moment map in the sense of [8].

2 L_∞-algebras

Let $\mathbb{K}$ denote here a fixed ground field of characteristic 0. All vector spaces, linear maps and tensor products will be defined with respect to/taken over this field, unless we explicitly state otherwise. A good overview of the subject of L_∞-algebras is provided in the n-lab ([12]).

2.1 Generalizing differential graded Lie algebras

In this section we will define L_∞-algebras as the generalization of Lie algebras. For doing so let us first review the notion of a Lie algebra: A Lie algebra is a vector space L with a skew-symmetric bilinear map $[\cdot,\cdot] : L \times L \to L$ satisfying the Jacobi identity:

$$[[x_1, x_2], x_3] - [[x_1, x_3], x_2] + [[x_2, x_3], x_1] = 0 \tag{1}$$

This can be rewritten in the following way, where $P = \{\left(\begin{smallmatrix} 1 & 2 & 3 \\ 1 & 2 & 3 \end{smallmatrix}\right), \left(\begin{smallmatrix} 1 & 2 & 3 \\ 1 & 3 & 2 \end{smallmatrix}\right), \left(\begin{smallmatrix} 1 & 2 & 3 \\ 2 & 3 & 1 \end{smallmatrix}\right)\} \subset S_3$:

$$\sum_{\sigma \in P} sgn(\sigma)[[x_{\sigma(1)}, x_{\sigma(2)}], x_{\sigma(3)}] = 0.$$

The set P consists precisely of those permutations, which move one element to the last position, without distorting the inner order of the others. As we will deal with identities of multi-brackets soon, we will have to generalize this notion of "moving one element out of three to the end" to "moving q elements out of $p + q$ to the end". The permutations doing that are exactly the (p, q)-unshuffles.

Definition 2.1. A permutation $\sigma \in S_{p+q}$ is a (p, q)-*unshuffle* if and only if $\sigma(i) < \sigma(i + 1)$ for $i \neq p$. We denote the set of of (p, q)-unshuffles by $ush(p, q) \subset S_{p+q}$.

The condition in the above definition guarantees that the first p and the last q elements stay in the same internal order. These permutations are called unshuffles, because their inverses correspond to shuffling a deck of p cards into a deck of q cards.

Let us now turn to the graded context. First we introduce a grading on our vector space: We set $L = \bigoplus_{i \in \mathbb{Z}} L_i$, where L_i ist the vector subspace of elements of degree i. We will write $|x| = i$ if $x \in L_i$. A Lie structure $[\cdot,\cdot]$ on such a vector space L should satisfy the following three conditions:

- $[L_i, L_j] \subset L_{i+j}$ (the bracket respects the grading)
- $[x_1, x_2] = -(-1)^{|x_1||x_2|}[x_2, x_1]$ for all homogenous $x_1, x_2 \in L$ (the bracket is graded skew-symmetric)
- $(-1)^{|x_1||x_3|}[x_1, [x_2, x_3]] + (-1)^{|x_1||x_2|}[x_2, [x_3, x_1]] + (-1)^{|x_2||x_3|}[x_3, [x_1, x_2]] = 0$ for all homogenous elements $x_1, x_2, x_3 \in L$ (the graded Jacobi identity holds)

If we try to bring the graded Jacobi identity into form of equation (1), we get:

$$[[x_1, x_2], x_3] - (-1)^{|x_2||x_3|}[[x_1, x_3], x_2] + (-1)^{|x_1||x_2|+|x_1||x_3|}[[x_2, x_3], x_1] = 0.$$

So $[[x_{\sigma(1)}, x_{\sigma(2)}], x_{\sigma(3)}]$ gets an additional sign for every transposition of two odd elements. This leads us to the definition of the Koszul sign ϵ of a permutation σ acting on elements $x_1, ..., x_n \in L$.

Definition 2.2. Let $\sigma \in S_n$ be a permutation acting on elements $v_1, ..., v_n$ of a $\mathbb{Z}$-graded vector space V. Let $(v_{i_1}, ..., v_{i_k})$ be the ordered sublist of $v_1, ..., v_n$ including exactly the odd elements. Then there is a (unique) permutation $\tilde{\sigma} \in S_k$ such that $(v_{i_{\tilde{\sigma}(1)}}, ..., v_{i_{\tilde{\sigma}(k)}})$ is the ordered sublist of $v_{\sigma(1)}, ..., v_{\sigma(n)}$ including exactly the odd elements. Then the *Koszul sign of σ acting on $v_1, ..., v_n$* is defined by

$$\epsilon(\sigma, v_1, ..., v_n) := sgn(\tilde{\sigma}).$$

Remark 2.3. One can check that ϵ is well-behaved in the sense that

$$\epsilon(\sigma' \circ \sigma, v_1, ..., v_k) = \epsilon(\sigma', v_{\sigma(1)}, ..., v_{\sigma(k)}) \cdot \epsilon(\sigma, v_1, ..., v_k),$$

and that for a transposition τ_i interchanging v_i and v_{i+1} it holds that $\epsilon(\tau_i, v_1, ..., v_n) = (-1)^{|v_i||v_{i+1}|}$.

Thus, the graded Jacobi identity can be written in the following way:

$$\sum_{\sigma \in ush(2,1)} sgn(\sigma)\epsilon(\sigma, x_1, x_2, x_3)[[x_{\sigma(1)}, x_{\sigma(2)}], x_{\sigma(3)}] = 0 \qquad (2)$$

Next we consider a graded vector space with a differential $d : L \to L$ satisfying $d(L_i) \subset L_{i+1}$ and $d^2 = 0$. This turns L into a differential graded vector space or, in other words, into a chain complex[1]:

$$\cdots \longrightarrow L_{i-2} \xrightarrow{d} L_{i-1} \xrightarrow{d} L_i \xrightarrow{d} L_{i+1} \xrightarrow{d} \cdots$$

Adopting the standard language, used e.g. for de Rham cohomology, we call $x \in L$ *closed* if $dx = d(x) = 0$ and *exact* if $x = dy$ for some $y \in L$. In the latter case y is called a *potential* for x.

A differential graded vector space (L, d) together with a graded Lie bracket $[\cdot, \cdot]$ on L is called a *differential graded Lie algebra* if the differential derives the bracket i.e. satisfies the following graded Leibniz rule (for $x_1, x_2 \in L$):

$$d[x_1, x_2] = [d(x_1), x_2] - (-1)^{|x_1|}[x_1, d(x_2)].$$

This can be rewritten as:

$$d[x_1, x_2] = [d(x_1), x_2] - (-1)^{|x_1||x_2|}[d(x_2), x_1].$$

The latter equation can also be written in terms of signed sums of unshuffles. In fact, it is equivalent to:

$$\sum_{\sigma \in ush(2,0)} sgn(\sigma)\epsilon(\sigma, x_1, x_2)d[x_{\sigma(1)}, x_{\sigma(2)}] = \sum_{\sigma \in ush(1,1)} sgn(\sigma)\epsilon(\sigma, x_1, x_2)[d(x_{\sigma(1)}), x_{\sigma(2)}].$$

[1]From the perspective of homological algebra it would be a cochain complex, as the differential has positive degree. The closed elements would be called cocycles and the exact elements coboundaries.

If we furthermore interpret the differential as a unary bracket and write $x \mapsto [x]$ instead of $x \mapsto d(x)$ we get:

$$\sum_{\sigma \in ush(2,0)} sgn(\sigma)\epsilon(\sigma, x_1, x_2)[[x_{\sigma(1)}, x_{\sigma(2)}]] = \sum_{\sigma \in ush(1,1)} sgn(\sigma)\epsilon(\sigma, x_1, x_2)[[x_{\sigma(1)}], x_{\sigma(2)}] \tag{3}$$

Even the condition $d^2 = 0$ can be written as $[[x]] = 0$, or bringing it into the form of the other equations :

$$\sum_{\sigma \in ush(1,0)} sgn(\sigma)\epsilon(\sigma, x_1)[[x_{\sigma(1)}]] = 0 \tag{4}$$

We have now learned how to describe a differential graded Lie algebra as a graded vector space L with a unary bracket $[\cdot]$ of degree one and a binary bracket $[\cdot,\cdot]$ of degree 0, which satisfy the equations (2), (3) and (4). These three equations are special cases of the below equation (5). We will now generalize the notion of differential graded Lie algebra to obtain a definition of an L_∞-algebra.

Definition 2.4. An L_∞*-algebra* (or *Lie-∞-algebra*) is a graded vector space $L = \bigoplus_{i \in \mathbb{Z}} L_i$ together with a family of graded skew-symmetric multilinear maps $\{l_i : \times^i L \to L \mid i \in \mathbb{N}\}$ such that l_i has degree $2-i$ and the following identity holds (for all $n \in \mathbb{N}$):

$$\sum_{i+j=n+1} (-1)^{i(j+1)} \sum_{\sigma \in ush(i,n-i)} sgn(\sigma)\epsilon(\sigma, x_1, ..., x_n)\, l_j(l_i(x_{\sigma(1)}, ..., x_{\sigma(i)}), x_{\sigma(i+1)} ..., x_{\sigma(n)}) = 0 \tag{5}$$

To keep the definition transparent despite the use of (possibly infinitely many) multi-brackets we write l_n for the n-ary bracket and give a brief overview of the multilinear algebra surpressed in the definition.

- $\times^i L$ is the cross product of i copies of L.
- A multi-linear map $l_i : \times^i L \to L$ is a map linear in every component. One could equivalently define l_i as a linear map from $\bigotimes^i L$ to L. Here $\bigotimes^i L$ is the i-fold tensor product of the graded vector space L.
- Demanding l_i to have degree $2-i$ means that l_i restricted to $L_{k_1} \times L_{k_2} \times ... \times L_{k_i}$ must map into $L_{k_1+k_2+...+k_i+2-i}$. Turning $\bigotimes^i L$ into a graded vector space by defining the degree of $x_1 \otimes ... \otimes x_i$ as $\sum_{k=1}^i |x_k|$, this translates to saying that $l_i : \bigotimes^i L \to L$ is a linear map of degree $2-i$.
- The maps l_i being (graded) skew-symmetric means that for all $\sigma \in S_i$ the identity $l_i(x_1, ..., x_i) = sgn(\sigma)\epsilon(\sigma, x_1, ..., x_i) l_i(x_{\sigma(1)}, ..., x_{\sigma(i)})$ holds. Using the language of multilinear algebra this is equivalent to $l_i : \bigotimes^i L \to L$ descending to a map $l_i : E^i(L) \to L$, where $E^i(L)$ is the i-th (graded) exterior power of L.

- The reason why we only sum over the unshuffles is to avoid repetition: If any two permutations $\sigma, \sigma' \in S_n$ differ by a permutation which only interchanges the first i and the last $n-i$ elements (i.e. $\sigma = \tau \circ \sigma'$, where $\tau = (\tau_1, \tau_2) \in S_i \times S_{n-i} \subset S_n$) then $l_j(l_i(x_{\sigma(1)}, ..., x_{\sigma(i)}), x_{\sigma(i+1)}...x_{\sigma(n)}) = \pm l_j(l_i(x_{\sigma'(1)}, ..., x_{\sigma'(i)}), x_{\sigma'(i+1)}...x_{\sigma'(n)})$. Thus, up to sign, we would add up essentially same elements several times. To avoid that, we need a system of representatives for $S_n/(S_i \times S_{n-i})$. That system of representatives is provided by the Unshuffles. All inner permutations of the first i and last $n - i$ elements are prohibited, as there is only one way to arrange an i-element (resp. $(n - i)$-element) subset of $\{1, ..., n\}$ in a strictly ascending order.

Next, let us take a closer look at the signs involved:

- The first sign in (5) is $(-1)^{i(j+1)}$. It only gives us an additional sign if the number of elements consumed by brackets is odd for the inner one (l_i) and even for the outer one (l_j). If we look at (5) for $n = 2$ we get equation (3). If not for the $(-1)^{i(j+1)}$ in our definition this equation would have an additional sign and the graded Leibniz rule would not be satisfied.

- $sgn(\sigma)$ is the usual sign of the permutation, not depending on the grading. It contributes a minus sign for every transposition in the permutation.

- $\epsilon(\sigma, x_1, ..., x_i)$ is the Koszul sign, which is highly dependant on the grading of the elements permuted. It gives us a factor of -1 for every transposition of two odd elements in the permutation.

Finally, by direct calculation, we can see that this definition indeed provides a generalization of a (differential graded) Lie algebra:

Example 2.5. A differential graded Lie algebra is precisely an L_∞-algebra with $l_i = 0$ for $i > 2$. Furthermore, a Lie algebra is an L_∞-algebra where the L is concentrated in degree zero (i.e. $L = L_0$).

2.2 L_∞-algebras as coalgebras with differentials

Having defined L_∞-algebra objects, we would now like to investigate their structure-preserving maps. Intuitively one would regard (say degree-zero) linear maps f which conserve the brackets i.e. $l_i'(f(x_1), ..., f(x_i)) = f(l_i(x_1, ..., x_i))$. Unfortunately this nontion of morphism is not flexible enough, and from our definition it is not clear what the right notion of more "flexible" maps would be. In order to advance, we will reformulate our definition in the language of "differential-graded coalgebras". First of all we will get rid of the different degrees of the l_n. We define the shift sL to be L with the grading shifted by one i.e. $(sL)_i = L_{i+1}$. To keep notation consistent we denote the multi-brackets by sl_i instead of l_i if we work with the shifted grading. Then we get

$$(sl_i)((sL)_{k_1}, ..., (sL)_{k_i}) = l_i(L_{k_1+1}, ..., L_{k_i+1}) \subset L_{k_1+1+...+k_i+1+2-i} = L_{k_1+...k_i+2} = (sL)_{k_1+...+k_i+1}.$$

So $(sl_i) : \bigtimes^i(sL) \to (sL)$ have degree one independantly of i. Unfortunately the sl_i are not in general skew-symmetric with respect to the new grading. But we can turn the l_i into graded

commutative maps by altering the signs.
Recall that a multi-linear map $f : \bigtimes^n V \to W$, where V is a graded vector space and W is any vector space, is called *graded-commutative* (or *graded-symmetric*) if it satisfies

$$f(y_1, ..., y_n) = \epsilon(\sigma, y_1, ..., y_n) f(y_{\sigma(1)}, ..., y_{\sigma(n)}) \quad \forall \sigma \in S_n \tag{6}$$

Writing $sx \in (sL)_n$ for an element $x \in L_{n+1}$, we now define $\hat{l}_n : \bigtimes^n(sL) \to (sL)$ by

$$\hat{l}_n(sx_1, ..., sx_n) = (-1)^{\alpha(x_1,...,x_n)}(sl_n)(sx_1, ..., sx_n)$$

with α defined in the following way:

$$\alpha(x_1, ..., x_n) = \alpha_n(x_1, ..., x_n) = \sum_{\substack{i \text{ is odd} \\ i \in \{1,...,n\}}} |x_i| \quad \text{when n is even,}$$

$$\alpha(x_1, ..., x_n) = \alpha_n(x_1, ..., x_n) = 1 + \sum_{\substack{i \text{ is even} \\ i \in \{1,...,n\}}} |x_i| \quad \text{when n is odd.}$$

Lemma 2.6. The maps $\hat{l}_n : \bigtimes^n(sL) \to sL$ are graded-symmetric multi-linear maps of degree 1 for all $n \in \mathbb{N}$.

Proof. The $\hat{l}_n$ are multi-linear by construction and $deg(\hat{l}_n) = 1$ follows from the discussion above. The only property, which remains to be checked is the graded symmetry (6). However, it is enough to check this property for the transpositions $\{\tau_i \mid 1 \leq i < n\} \subset S_n$, where τ_i interchanges y_i and y_{i+1}, as these transpositions generate the full symmetric group. So we calculate:

$$\begin{aligned}\hat{l}_n(sx_1, ..., sx_n) &= (-1)^{\alpha(x_1,...,x_n)} sl_n(sx_1, ..., sx_n) = (-1)^{\alpha(x_1,...,x_n)} l_n(x_1, ..., x_n) \\ &= (-1)^{\alpha(x_1,...,x_n)} sgn(\tau_i)\epsilon(\tau_i, x_1, ..., x_n) l_n(x_{\tau_i(1)}, ..., x_{\tau_i(n)}).\end{aligned}$$

As τ_i consists of only one transposition we have $sgn(\tau_i) = -1$ and $\epsilon(\tau_i, x_1, ..., x_n) = (-1)^{|x_i||x_{i+1}|}$, thus

$$\begin{aligned}\hat{l}_n(sx_1, ..., sx_n) &= (-1)^{\alpha(x_1,...,x_n)}(-1)(-1)^{|x_i||x_{i+1}|} l_n(x_{\tau_i(1)}, ..., x_{\tau_i(n)}) \\ &= (-1)^{\alpha(x_1,...,x_n)}(-1)(-1)^{|x_i||x_{i+1}|} sl_n(sx_{\tau_i(1)}, ..., sx_{\tau_i(n)}) \\ &= (-1)^{\alpha(x_1,...,x_n)}(-1)(-1)^{|x_i||x_{i+1}|}(-1)^{\alpha(x_{\tau_i(1)},...,x_{\tau_i(n)})} \hat{l}_n(sx_{\tau_i(1)}, ..., sx_{\tau_i(n)}).\end{aligned}$$

But the sum $\alpha(x_1, ..., x_n)$ and $\alpha(x_{\tau_i(1)}, ..., x_{\tau_i(n)})$ only differ by one summand. One of them has the summand $|x_i|$ and the other one $|x_{i+1}|$. All the other signs cancel out, so we get:

$$\begin{aligned}\hat{l}_n(sx_1, ..., sx_n) &= (-1)^{|x_i|+|x_j|}(-1)(-1)^{|x_i||x_{i+1}|} \hat{l}_n(sx_{\tau_i(1)}, ..., sx_{\tau_i(n)}) = (-1)^{(|x_i|+1)(|x_{i+1}|+1)} \hat{l}_n(sx_{\tau_i(1)}, ..., sx_{\tau_i(n)}) \\ &= \epsilon(\tau_i, sx_1, ..., sx_n) \hat{l}_n(sx_{\tau_i(1)}, ..., sx_{\tau_i(n)}).\end{aligned}$$

□

Remark 2.7. In this proof the constant summand in the definition of α for odd n is not needed. It will turn out useful later, when we discuss the multi-bracket equation (5).

We can now regard $\hat{l}_n$ as a graded-symmetric linear map from $\bigotimes^n(sL)$ to sL of degree one. And, analogously, we can encode the symmetry properties of this linear map by modifying its domain. This time we need the graded-symmetric powers of (sL), which are denoted by $S^n(sL)$. Analogusly to the preceeding reasoning we can describe $\hat{l}_n$ equivalently as linear maps $\hat{l}_n : S^n(sL) \to sL$, where $S^n(\cdot)$ is defined as follows:

Definition 2.8. Let V be a $\mathbb{Z}$-graded vector space. *The k-th (graded) symmetrization operator* $Sym_k : \bigotimes^k V \to \bigotimes^k V$ is defined by

$$Sym_k(v_1 \otimes ... \otimes v_k) := \frac{1}{k!} \sum_{\sigma \in S_k} \epsilon(\sigma, v_1, ..., v_k) v_{\sigma(1)} \otimes ... \otimes v_{\sigma(k)}.$$

The Image of Sym_k is a vector subspace and called *the k-th (graded) symmetric power of* V. We will denote $Sym_k(v_1 \otimes ... \otimes v_k)$ by $v_1 \odot ... \odot v_k$.

The sum of the $S^n(sL)$ forms an algebra $S^\bullet(sL) = \bigoplus_{n \in \mathbb{N}_0} S^n(sL)$, the so-called *free graded-commutative algebra* on sL. Setting $\hat{l}_0 : S^0(sL) = \mathbb{K} \to (sL)$ to be the zero map, we can combine the $\hat{l}_i$ to a linear map $\hat{l} : S^\bullet(sL) \to sL$ of degree 1. This map encodes all information of our L_∞-algebra except equation (5).

To encode equation (5) we have to change perspectives. Instead of regarding $S^\bullet(sL)$ as an algebra we will use its coalgebra structure. The reason for this is that the universal property of an algebra gives us a way to extend the domain of maps $sL \to W$. What we need here is the dual property: We want to extend a map $W \to sL$ to some map $W \to S^\bullet(sL)$ (in our case $W = S^\bullet(sL)$ and the map to be extended is $\hat{l}$). The most direct way to understand a coalgebra is by dualizing the diagrams encoding an algebra structure. Let us first recall the definition of a graded algebra in terms of diagrams:

Definition 2.9. Let A be a graded vector space. A *unital associative graded commutative algebra-structure on* A consists of two linear maps of degree zero, a linear multiplication map $m : A \otimes A \to A$ and a unit morphism $u : \mathbb{K} \to A$ such that the following diagrams commute, where $\tau : A \otimes A \to A \otimes A$ is defined by $\tau(a \otimes b) = (-1)^{|a||b|} b \otimes a$:

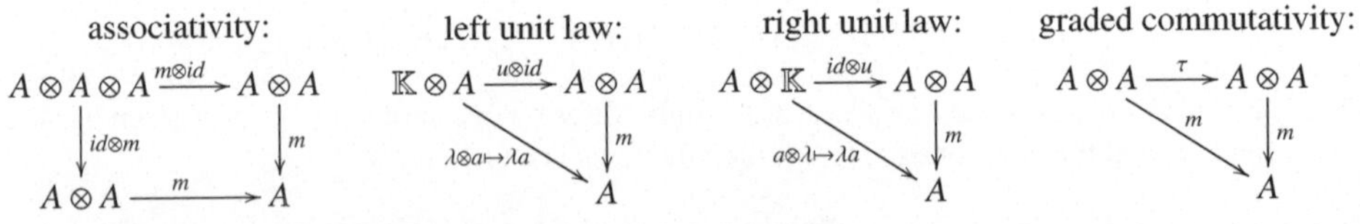

"Reversing all arrows" gives us the definition of a coalgebra:

Definition 2.10. Let C be a graded vector space. A *counital coassociative graded cocommutative coalgebra-structure on* C consists of two linear maps of degree zero, a comultiplication map

$\Delta : C \otimes C \to C$ (also called diagonal) and a counit morphism $\pi : C \to \mathbb{K}$ such that the following diagrams commute:

coassociativity:

$$
\begin{array}{ccc}
C & \xrightarrow{\Delta} & C \otimes C \\
\downarrow{\scriptstyle \Delta} & & \downarrow{\scriptstyle \Delta \otimes id} \\
C \otimes C & \xrightarrow{id \otimes \Delta} & C \otimes C \otimes C
\end{array}
$$

left counit law:

$$
\begin{array}{ccc}
C & \xrightarrow{\Delta} & C \otimes C \\
{\scriptstyle c \mapsto 1 \otimes c}\downarrow & & \downarrow{\scriptstyle \pi \otimes id} \\
 & \mathbb{K} \otimes C &
\end{array}
$$

right counit law:

$$
\begin{array}{ccc}
C & \xrightarrow{\Delta} & C \otimes C \\
{\scriptstyle c \mapsto c \otimes 1}\downarrow & & \downarrow{\scriptstyle id \otimes \pi} \\
 & \mathbb{K} \otimes C &
\end{array}
$$

graded cocommutativity:

$$
\begin{array}{ccc}
 & C & \\
{\scriptstyle \Delta}\swarrow & & \searrow{\scriptstyle \Delta} \\
C \otimes C & \xrightarrow{\tau} & C \otimes C
\end{array}
$$

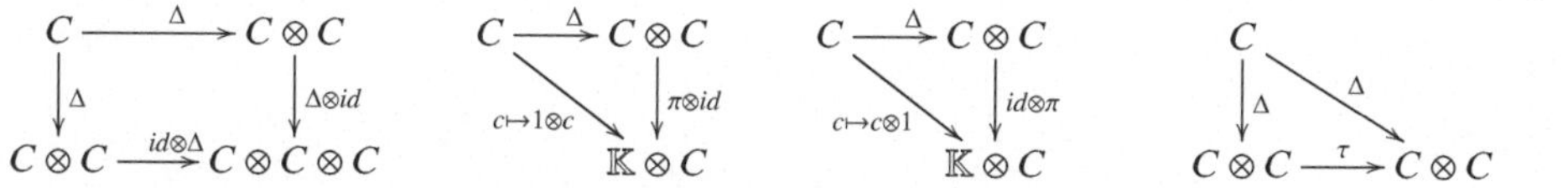

One can turn $S^\bullet(sL)$ into a counital coassociative graded cocommutative coalgebra with the counit $\pi = \pi_0$ given by the projection onto $S^0(sL) = \mathbb{K}$ and the following diagonal:

$$\Delta(sx_1 \odot ... \odot sx_n) := \sum_{i=0}^{n} \sum_{\sigma \in ush(i,n-i)} \epsilon(\sigma, sx_1, ..., sx_n) \left(sx_{\sigma(1)} \odot ... \odot sx_{\sigma(i)} \right) \otimes \left(sx_{\sigma(i+1)} \odot ... \odot sx_{\sigma(n)} \right)$$

More generally, this construction can be carried out for any graded vector space V. The resulting coalgebra structure on the graded vector space $S^\bullet V$ is then called the *cofree graded cocommutative coalgebra* of V. For any counital coassociative algebra C a homomorphism $\delta : C \to C$ is called a *coderivation* if the following diagram commutes:

$$
\begin{array}{ccc}
C & \xrightarrow{\delta} & C \\
\downarrow{\scriptstyle \Delta} & & \downarrow{\scriptstyle \Delta} \\
C \otimes C & \xrightarrow{a \otimes b \mapsto \delta(a) \otimes b + (-1)^{|\delta||a|} a \otimes \delta(b)} & C \otimes C
\end{array}
$$

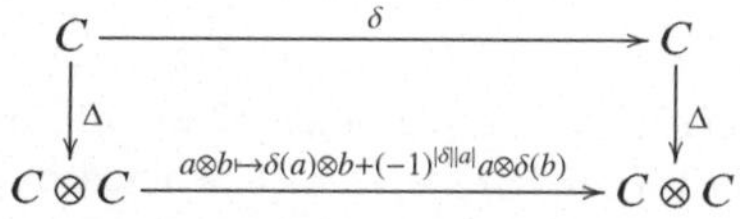

Returning to our case, we have a map $\hat{l} : S^\bullet(sL) \to sL = S^1(sL)$. It can be extended to a degree 1 coderivation $D : S^\bullet(sL) \to S^\bullet(sL)$ in the following way:

Definition 2.11. Let L be a $\mathbb{Z}$-graded vector space and $\{l_i \mid i \in \mathbb{N}\}$ a family of graded skew-symmetric maps. Let $\hat{l} : S^\bullet(sL) \to sL$ be the above defined map. Then we define the degree-one derivation $D : S^\bullet(sL) \to S^\bullet(sL)$ by the following formula:

$$D(sx_1 \odot ... \odot sx_n) = \sum_{i=1}^{n} \sum_{\sigma \in ush(i,n-i)} \epsilon(\sigma, sx_1, ..., sx_n) \left(\hat{l}(sx_{\sigma(1)} \odot ... \odot sx_{\sigma(i)}) \right) \odot sx_{\sigma(i+1)} \odot ... \odot sx_{\sigma(n)}.$$

Remark 2.12. As we defined $\hat{l}$ summand-wise by the $\hat{l}_i$, we could instead write

$$D(sx_1 \odot ... \odot sx_n) = \sum_{i=1}^{n} \sum_{\sigma \in ush(i,n-i)} \epsilon(\sigma, sx_1, ..., sx_n) \left(\hat{l}_i(sx_{\sigma(1)} \odot ... \odot sx_{\sigma(i)}) \right) \odot sx_{\sigma(i+1)} \odot ... \odot sx_{\sigma(n)},$$

and check that D indeed is a coderivation. As a conceqence of corollary A.2 D is the unique coderivation satisfying $\pi_1 D = \hat{l}$.

With this extension done, we can finally reformulate condition (5).

Lemma 2.13. In the setting of the last definition the equation

$$\sum_{i+j=n+1} (-1)^{i(j+1)} \sum_{\sigma \in ush(i,n-i)} sgn(\sigma)\epsilon(\sigma, x_1, ..., x_n)\, l_j(l_i(x_{\sigma(1)}, ..., x_{\sigma(i)}), x_{\sigma(i+1)}..., x_{\sigma(n)}) = 0$$

implies $D^2 = 0$.

Proof.

$$D^2(sx_1 \odot ... \odot sx_n) = D\left(\sum_{i=1}^{n} \sum_{\sigma \in ush(i,n-i)} \epsilon(\sigma, sx_1, ..., sx_n)\left(\hat{l}_i(sx_{\sigma(1)} \odot ... \odot sx_{\sigma(i)})\right) \odot sx_{\sigma(i+1)} \odot ... \odot sx_{\sigma(n)}\right)$$
$$= \sum_{i=1}^{n} \sum_{\sigma \in ush(i,n-i)} \epsilon(\sigma, sx_1, ..., sx_n) D\left(\left(\hat{l}_i(sx_{\sigma(1)} \odot ... \odot sx_{\sigma(i)})\right) \odot sx_{\sigma(i+1)} \odot ... \odot sx_{\sigma(n)}\right)$$

Next we apply D to a term of length $n-i+1$. For that we will define $y_1^\sigma := \hat{l}_i(sx_{\sigma(1)} \odot ... \odot sx_{\sigma(i)})$ and $y_{k+1}^\sigma := sx_{\sigma(i+k)}$ for $k \in \{1, ..., n-i\}$. So we have:

$$D\left(\left(\hat{l}_i(sx_{\sigma(1)} \odot ... \odot sx_{\sigma(i)})\right) \odot sx_{\sigma(i+1)} \odot ... \odot sx_{\sigma(n)}\right) = D(y_1^\sigma \odot ... \odot y_{n-i+1}^\sigma)$$
$$= \sum_{j=1}^{n-i+1} \sum_{\tau \in ush(j,n-i+1-j)} \epsilon(\tau, y_1^\sigma, ..., y_{n-i+1}^\sigma)\hat{l}_j(y_{\tau(1)}^\sigma \odot ... \odot y_{\tau(j)}^\sigma) \odot y_{\tau(j+1)}^\sigma \odot ... \odot y_{\tau(n-i+1)}^\sigma.$$

As y_1^σ is a structurally different term than the other y_k^σ we will distinguish between τ satisfying $\tau(1) = 1$ denoted by $u_1(j, n, i)$ and $\tau(j+1) = 1$ denoted by $u_2(j, n, i)$. Every element in $ush(j, n-i+1-j)$ is exactly in one of those two subsets. Let us first analyse the case $\tau \in u_2(j, n, i)$:

$$\hat{l}_j(y_{\tau(1)}^\sigma \odot ... \odot y_{\tau(j)}^\sigma) \odot y_{\tau(j+1)}^\sigma \odot ... \odot y_{\tau(n-i+1)}^\sigma$$
$$= \hat{l}_j(y_{\tau(1)}^\sigma \odot ... \odot y_{\tau(j)}^\sigma) \odot \hat{l}_i(sx_{\sigma(1)} \odot ... \odot sx_{\sigma(i)}) \odot y_{\tau(j+2)}^\sigma \odot ... \odot y_{\tau(n-i+1)}^\sigma.$$

The total sign of this element is $\epsilon(\sigma, sx_1, ..., sx_n) \cdot \epsilon(\tau, y_1^\sigma, ..., y_{n-i+1}^\sigma)$. Now we regard the summand coming from $\tilde{i} = j, \tilde{j} = i$ and $\tilde{\sigma} \in ush(\tilde{i}, n-\tilde{i})$, $\tilde{\tau} \in u_2(\tilde{j}, n, \tilde{i})$ defined as follows:

- $\tilde{\sigma}$ is the unshuffle that, given the strictly ascending list $(sx_1, ..., sx_n)$, moves the elements $y_{\tau(1)}^\sigma, ..., y_{\tau(j)}^\sigma$ to the front.
- $\tilde{\tau}$ is the unshuffle that, given a list starting with $\hat{l}_{\tilde{i}}(y_{\tau(1)}^\sigma \odot ... \odot y_{\tau(j)}^\sigma)$ and continuing with a strictly ascending list of $\left\{y_{\tau(j+2)}^\sigma, ..., y_{\tau(n-i+1)}^\sigma\right\} \cup \left\{sx_{\sigma(1)}, ..., sx_{\sigma(i)}\right\}$, moves the elements $(sx_{\sigma(1)}, ..., sx_{\sigma(i)})$ to the front.

By construction the summands belonging to (i, j, σ, τ) and $(\tilde{i}, \tilde{j}, \tilde{\sigma}, \tau)$ are equal up to sign. The signs come from the odd transpositions in $\sigma, \tau, \tilde{\sigma}, \tilde{\tau}$ and from interchanging $\hat{l}_i(sx_{\sigma(1)} \odot ... \odot sx_{\sigma(i)})$ with $\hat{l}_j(y_{\tau(1)}^\sigma \odot ... \odot y_{\tau(j)}^\sigma)$.

First of all let us take a look at the transpositions nessecary to move the elements $y_{\tau(k)}^\sigma$ with

$k \in \{j+2, ..., n-i+1\}$ to the end. So we fix such a k. In the summand belonging to (σ, τ), there is some subset A of $\{sx_{\sigma(1)}, ... sx_{\sigma(i)}\}$ which has to be moved to the front past $y^{\sigma}_{\tau(k)}$, while applying σ. After that operation there is a subset B of $\{y^{\sigma}_{\tau(1)} ... y^{\sigma}_{\tau(j)}\}$ which has to be moved to the front past $y^{\sigma}_{\tau(k)}$. In the summand belonging to $(\tilde{\sigma}, \tilde{\tau})$ by construction first the elements of B and then the elements of A have to be moved past $y^{\sigma}_{\tau(k)}$. So $y^{\sigma}_{\tau(k)}$ is part of the same transpositions in both summands.

Next let us consider the transpositions of $sx_{\sigma(k)}$ for $k \in \{1, ..., i\}$ while applying σ. For fixed k $x_{\sigma(k)}$ has to be moved past some elements $A_k \subset \{y^{\sigma}_{\tau(1)}, ..., y^{\sigma}_{\tau(j)}\}$. Then in the course of applying τ $\hat{l}_i(sx_{\sigma(1)} \odot ... \odot sx_{\sigma(i)})$ has to be moved past all $\{y^{\sigma}_{\tau(1)}, ..., y^{\sigma}_{\tau(j)}\}$. Analogously for $l \in \{1, ..., j\}$ while applying $\tilde{\sigma}$ we have to move $y^{\sigma}_{\tau(l)}$ past $B_l \subset \{sx_{\sigma(1)}, ..., \sigma(i)\}$. Also we have to move $\hat{l}_j(y^{\sigma}_{\tau(1)} \odot ... \odot y^{\sigma}_{\tau(j)})$ past all $\{sx_{\sigma(1)}, ..., sx_{\sigma(i)}\}$ while applying $\tilde{\tau}$. Note that,by construction, it holds that

$$\bigcup_{k \in \{1,...,i\}} (\{k\} \times A_k) \cup \bigcup_{l \in \{1,...,j\}} (B_l \times \{l\}) = \{1, ..., i\} \times \{1, ..., j\} \tag{7}$$

Note that $(\{k\} \times A_k) \cap (B_l \times \{l\}) = \emptyset$ for all k and l such that all unions involved are disjoint. Thus the relative sign of the two summands is $(-1)^c$ for

$$c = \sum_{k=1}^{i} \sum_{l \in A_k} |sx_{\sigma(k)}||y^{\sigma}_{\tau(l)}| + \left(1 + \sum_{k=1}^{i} |sx_{\sigma(k)}|\right) \sum_{l=1}^{j} |y^{\sigma}_{\tau(l)}| + \sum_{l=1}^{j} \sum_{k \in B_l} |sx_{\sigma(k)}||y^{\sigma}_{\tau(l)}|$$
$$+ \sum_{k=1}^{i} |sx_{\sigma(k)}| \left(1 + \sum_{l=1}^{j} |y^{\sigma}_{\tau(l)}|\right) + \left(1 + \sum_{k=1}^{i} |sx_{\sigma(k)}|\right)\left(1 + \sum_{l=1}^{j} |y^{\sigma}_{\tau(l)}|\right)$$

The first summand comes from permuting $sx_{\sigma(k)}$ with $y^{\sigma}_{\tau(l)}$ via σ, the second one from permuting $\hat{l}_i(...)$ with $y^{\sigma}_{\tau(l)}$ via τ, the third one from permuting $y^{\sigma}_{\tau(l)}$ with $sx_{\sigma(k)}$ via $\tilde{\sigma}$, the fourth one from permuting $\hat{l}_j(...)$ with $sx_{\sigma(k)}$ and the last summand comes from interchanging $\hat{l}_i(...)$ with $\hat{l}_j(...)$. As already discussed, all other transpositions (i.e. those with $y^{\sigma}_{\tau(l)}$ for $l \geq j+2$) happen in both summands and thus give no relative sign. Using (7) we observe that:

$$\sum_{k=1}^{i} \sum_{l \in A_k} |sx_{\sigma(k)}||y^{\sigma}_{\tau(l)}| + \sum_{l=1}^{j} \sum_{k \in B_l} |sx_{\sigma(k)}||y^{\sigma}_{\tau(l)}| = \sum_{k=1}^{i} \sum_{l=1}^{j} |sx_{\sigma(k)}||y^{\sigma}_{\tau(l)}| = \left(\sum_{k=1}^{i} |sx_{\sigma(k)}|\right)\left(\sum_{l=1}^{j} |y^{\sigma}_{\tau(l)}|\right)$$

Writing p for $(\sum_{k=1}^{i} |sx_{\sigma(k)}|$ and q for $\sum_{l=1}^{j} |y^{\sigma}_{\tau(l)}|$ we get:

$$c = pq + (1+p)q + p(1+q) + (1+p)(1+q) = (1+p)(1+q) - 1 + (1+p)(1+q) \equiv 1 \mod 2$$

Thus the relative sign is -1 and we have shown, that summands coming from $u_2(j, n, i)$ cancel out pairwise.

So (without using equation (5)) we have shown:

$$D^2(sx_1 \odot ... \odot sx_n) = \sum_{i=1}^{n} \sum_{\sigma \in ush(i,n-i)} \epsilon(\sigma, sx_1, ..., sx_n) \sum_{j=1}^{n-i+1} \sum_{\tau \in u_1(j,n,i)} \epsilon(\tau, y_1^\sigma, ..., y_{n-i+1}^\sigma) \times \tag{8}$$

$$\hat{l}_j(\hat{l}_i(sx_{\sigma(1)} \odot ... \odot sx_{\sigma(i)}) \odot y^\sigma_{\tau(2)} \odot ... \odot y^\sigma_{\tau(j)}) \odot y^\sigma_{\tau(j+1)} \odot ... \odot y^\sigma_{\tau(n-i+1)}$$

$$= \sum_{\substack{i \in \{1,...,n\} \\ \sigma \in ush(i,n-i) \\ j \in \{1,...,n-i+1\} \\ \tau \in u_1(j,n,i)}} \epsilon(\sigma, sx_1, ..., sx_n)\epsilon(\tau, y_1^\sigma, ..., y_{n-i+1}^\sigma) \times$$

$$\hat{l}_j(\hat{l}_i(sx_{\sigma(1)} \odot ... \odot sx_{\sigma(i)}) \odot y^\sigma_{\tau(2)} \odot ... \odot y^\sigma_{\tau(j)}) \odot y^\sigma_{\tau(j+1)} \odot ... \odot y^\sigma_{\tau(n-i+1)}.$$

As expressions of different length can not cancel out each other, we can group elements of some fixed length d and thus obtain:

$$D^2(sx_1 \odot ... \odot sx_n) = \sum_{\substack{d \in \{1,...,n\} \\ i+j=d+1 \\ \sigma \in ush(i,n-i) \\ \tau \in u_1(j,n,i)}} \epsilon(\sigma, sx_1, ..., sx_n)\epsilon(\tau, y_1^\sigma, ..., y_{n-i+1}^\sigma) \times$$

$$\hat{l}_j(\hat{l}_i(sx_{\sigma(1)} \odot ... \odot sx_{\sigma(i)}) \odot y^\sigma_{\tau(2)} \odot ... \odot y^\sigma_{\tau(j)}) \odot y^\sigma_{\tau(j+1)} \odot ... \odot y^\sigma_{\tau(n-i+1)}.$$

The last sum is zero if and only if it is zero for any fixed value of d. So it suffices to analyse the following sums for $d \in \{1, ..., n\}$:

$$\sum_{\substack{i+j=d+1 \\ \sigma \in ush(i,n-i) \\ \tau \in u_1(j,n,i)}} \epsilon(\sigma, sx_1, ..., sx_n)\epsilon(\tau, y_1^\sigma, ..., y_{n-i+1}^\sigma) y^\sigma_{\tau(j)}) \times \tag{9}$$

$$\hat{l}_j(\hat{l}_i(sx_{\sigma(1)} \odot ... \odot sx_{\sigma(i)}) \odot y^\sigma_{\tau(2)} \odot ... \odot \odot y^\sigma_{\tau(j+1)} \odot ... \odot y^\sigma_{\tau(n-i+1)}$$

Now we have to regroup the permutations. Instead of moving $sx_{\sigma(1)} \odot ... \odot sx_{\sigma(i)}$ to the front and then $y^\sigma_{\tau(j+1)} \odot ... \odot y^\sigma_{\tau(n-i+1)}$ to the back, we first move $y^\sigma_{\tau(j+1)} \odot ... \odot y^\sigma_{\tau(n-i+1)}$ to the back via $\tilde{\tau} \in ush(d, n-d)$. Then we rename the elements $sx_{\tilde{\tau}(k)}$ with $k \in \{1, ..., d\}$ to $z_k^{\tilde{\tau}}$ and move $sx_{\sigma(1)} \odot ..., \odot sx_{\sigma(i)}$ to the front via some $\tilde{\sigma} \in ush\{i, d-i\}$. As we perform the same transpositions as before we can write the sign in terms of $\tilde{\tau}$ and $\tilde{\sigma}$. So (for each fixed d) the above sum is equal to:

$$\sum_{\substack{i+j=d+1 \\ \tilde{\sigma} \in ush(i,d-i) \\ \tilde{\tau} \in ush(d,n-d)}} \epsilon(\tilde{\sigma}, z_1^{\tilde{\tau}}, ... z_d^{\tilde{\tau}})\epsilon(\tilde{\tau}, sx_1, ... sx_n)\hat{l}_j(\hat{l}_i(z^{\tilde{\tau}}_{\tilde{\sigma}(1)} \odot ... \odot z^{\tilde{\tau}}_{\tilde{\sigma}(i)}) \odot z^{\tilde{\tau}}_{\tilde{\sigma}(i+1)} \odot ... \odot z^{\tilde{\tau}}_{\tilde{\sigma}(d)}) \odot sx_{\tilde{\tau}(d+1)} \odot ... \odot sx_{\tilde{\tau}(n)}$$

$$= \sum_{\tilde{\tau} \in ush(d,n-d)} \left(\sum_{\substack{i+j=d+1 \\ \tilde{\sigma} \in ush(i,d-i)}} \epsilon(\tilde{\sigma}, z_1^{\tau}, ... z_d^{\tau})\hat{l}_j(\hat{l}_i(z^{\tilde{\tau}}_{\tilde{\sigma}(1)} \odot ... \odot z^{\tilde{\tau}}_{\tilde{\sigma}(i)}) \odot z^{\tilde{\tau}}_{\tilde{\sigma}(i+1)} \odot ... \odot z^{\tilde{\tau}}_{\tilde{\sigma}(d)}) \right.$$

$$\left. \odot \epsilon(\tilde{\tau}, sx_1, ... sx_n) sx_{\tilde{\tau}(d+1)} \odot ... \odot sx_{\tilde{\tau}(n)}. \right.$$

Thus, it is sufficient to show, that the inner sum is zero for any fixed $\tilde{\tau}$. Terefore it only remains to show, that for any $sw_1, ..., sw_d \in sL$ the following expression is zero:

$$\sum_{\substack{i+j=d+1 \\ \sigma \in ush(i,d-i)}} \epsilon(\sigma, sw_1, ...sw_d)\hat{l}_j(\hat{l}_i(sw_{\sigma(1)} \odot ... \odot sw_{\sigma(i)}) \odot sw_{\sigma(i+1)} \odot ... \odot sw_{\sigma(d)}).$$

Resubstituting l_i and l_j for $\hat{l}_i$ and $\hat{l}_j$ we obtain for this sum:

$$\sum_{\substack{i+j=d+1 \\ \tilde{\sigma} \in ush(i,d-i)}} (-1)^{\alpha(w_{\sigma(1)},...,w_{\sigma(i)})}\epsilon(\sigma, sw_1, ...sw_d)\hat{l}_j(sl_i(w_{\sigma(1)} \odot ... \odot w_{\sigma(i)}) \odot sw_{\sigma(i+1)} \odot ... \odot sw_{\sigma(d)})$$

$$= \sum_{\substack{i+j=d+1 \\ \tilde{\sigma} \in ush(i,d-i)}} (-1)^{\beta}\epsilon(\sigma, sw_1, ...sw_d)sl_j(l_i(w_{\sigma(1)} \odot ... \odot w_{\sigma(i)}) \odot w_{\sigma(i+1)} \odot ... \odot w_{\sigma(d)})$$

$$= s\left(\sum_{\substack{i+j=d+1 \\ \tilde{\sigma} \in ush(i,d-i)}} (-1)^{\beta}\epsilon(\sigma, sw_1, ...sw_d)l_j(l_i(w_{\sigma(1)} \odot ... \odot w_{\sigma(i)}) \odot w_{\sigma(i+1)} \odot ... \odot w_{\sigma(d)})\right),$$

where $\beta = \alpha(w_{\sigma(1)}, ..., w_{\sigma(i)}) + \alpha(l_i(w_{\sigma(1)}, ..., w_{\sigma(i)}), w_{\sigma(i+1)}, ..., w_{\sigma(d)})$.

To finish the proof, we show that the following equation holds, where the additional sign $(\pm)$ only depends on $w_1, ..., w_d$:

$$(-1)^{\beta}\epsilon(\sigma, sw_1, ...sw_d) = \pm(-1)^{i(j+1)}sgn(\sigma)\epsilon(\sigma, w_1, ..., w_d)$$

First of all we will show, that the equation holds for the identity permutation.Let us consider some fixed $w_1, ..., w_d$ and $j = 1$. This will give us some definite sign instead of $\pm$. Next we will iteratively increment j (and thus automatically decrement i). In order to achieve that we will have to distinguish the following cases. In the subsequent calculations we use the definition of α and the fact, that l_k has degree $2-k$. We distinguish the cases arising through the parities of i and j:

- i and j are even: then $(-1)^{i(j+1)}$ gives us no relative sign and we calculate:

$$\beta \equiv \beta(i, j) \equiv \sum_{\substack{k \text{ is odd} \\ k \in \{1,...,i\}}} |w_k| + \left(2 - i + \sum_{k \in \{1,...,i\}} |w_k|\right) + \sum_{\substack{k \text{ is even} \\ k \in \{i+1,...,d\}}} |w_k|$$
$$\equiv -i + \sum_{\substack{k \text{ is even} \\ k \in \{1,...,i-1\}}} |w_k| + \sum_{\substack{k \text{ is even} \\ k \in \{i+1,...,d\}}} |w_k| \equiv -i + \sum_{\substack{k \text{ is even} \\ k \in \{1,...,d\}}} |w_k| \mod 2.$$

$$\beta(i-1, j+1) \equiv \left(1 + \sum_{\substack{k \text{ is even} \\ k \in \{1,...,i-1\}}} |w_k|\right) + \left(1 + \sum_{\substack{k \text{ is even} \\ k \in \{i,...,d\}}} |w_k|\right) \equiv \sum_{\substack{k \text{ is even} \\ k \in \{i,...,d\}}} |w_k| \mod 2.$$

As i is even this means we get no relative sign here either, so the signs are compatible.

- i and j are odd: then $(-1)^{i(j+1)}$ gives us no relative sign and we calculate:

$$\beta(i,j) \equiv \left(1 + \sum_{\substack{k \text{ is even} \\ k\in\{1,\dots,i\}}} |w_k|\right) + \left(1 + \sum_{\substack{k \text{ is even} \\ k\in\{i+1,\dots,d\}}} |w_k|\right) \equiv \sum_{\substack{k \text{ is even} \\ k\in\{i,\dots,d\}}} |w_k| \mod 2.$$

$$\begin{aligned}\beta(i-1,j+1) &\equiv \sum_{\substack{k \text{ is odd} \\ k\in\{1,\dots,i-1\}}} |w_k| + \left(2-(i-1) + \sum_{k\in\{1,\dots,i-1\}} |w_k|\right) + \sum_{\substack{k \text{ is even} \\ k\in\{i,\dots,d\}}} |w_k| \\ &\equiv 2-(i-1) + \sum_{\substack{k \text{ is even} \\ k\in\{1,\dots,i-1\}}} |w_k| + \sum_{\substack{k \text{ is even} \\ k\in\{i,\dots,d\}}} |w_k| \equiv -(i-1) + \sum_{\substack{k \text{ is even} \\ k\in\{1,\dots,d\}}} |w_k| \mod 2.\end{aligned}$$

  As $i-1$ is even this means we get no relative sign here either, so the signs are compatible.

- i is even and j is odd: $(-1)^{i(j+1)}$ gives us a relative sign and we calculate:

$$\beta(i,j) \equiv \sum_{\substack{k \text{ is odd} \\ k\in\{1,\dots,i\}}} |w_k| + 1 + \sum_{\substack{k \text{ is odd} \\ k\in\{i+1,\dots,d\}}} |w_k| \equiv 1 + \sum_{\substack{k \text{ is odd} \\ k\in\{1,\dots,d\}}} |w_k| \mod 2.$$

$$\begin{aligned}\beta(i-1,j+1) &\equiv 1 + \sum_{\substack{k \text{ is even} \\ k\in\{1,\dots,i-1\}}} |w_k| + \left(2-(i-1) + \sum_{k\in\{1,\dots,i-1\}} |w_k|\right) + \sum_{\substack{k \text{ is odd} \\ k\in\{i,\dots,d\}}} |w_k| \\ &\equiv -i + \sum_{\substack{k \text{ is odd} \\ k\in\{1,\dots,d\}}} |w_k| \mod 2.\end{aligned}$$

  As i is even this gives us a relative sign too, so the signs are compatible.

- i is odd and j is even: $(-1)^{i(j+1)}$ gives us a relative sign and we calculate:

$$\beta(i,j) \equiv 1 + \sum_{\substack{k \text{ is even} \\ k\in\{1,\dots,i\}}} |w_k| + \left((2-i) + \sum_{k\in\{1,\dots,i\}} |w_k|\right) + \sum_{\substack{k \text{ is odd} \\ k\in\{i+1,\dots,d\}}} |w_k| \equiv 1 - i + \sum_{\substack{k \text{ is odd} \\ k\in\{1,\dots,d\}}} |w_k| \mod 2.$$

$$\beta(i-1,j+1) \equiv \sum_{\substack{k \text{ is odd} \\ k\in\{1,\dots,i-1\}}} |w_k| + 1 + \sum_{\substack{k \text{ is odd} \\ k\in\{i,\dots,d\}}} |w_k| \equiv 1 + \sum_{\substack{k \text{ is odd} \\ k\in\{1,\dots,d\}}} |w_k| \mod 2.$$

  As $i-1$ is even this gives us a relative sign too, so the signs are compatible.

Thus we have shown, that the signs are compatible for all i, j and the identity permutation. As the transpositions generate the symmetric group, we just have to show that the formula holds true, when σ is a transposition. Lemma 2.6 already shows, that single brackets are compatible with transpositions, so we only have to check the transposition which moves an element from one bracket to the other. This is the transposition τ interchanging the i-th and the $(i+1)$-th element. Let us have a look at the signs involved:

- $sgn(\tau) = -1$
- $\epsilon(\tau, w_1, ..., w_d) = (-1)^{|w_i||w_{i+1}|}$
- $\epsilon(\tau, sw_1, ..., sw_d) = (-1)^{(|w_i|+1)(|w_{i+1}|+1)}$
- As we can see from the above calculations for fixed i and j, β either only sees the odd numered elements or only the even numbered elements. So by switching w_i and w_{i+1} exactly the other one of the two is "visible" and we get an additional sign $(-1)^{|w_i|+|w_{i+1}|}$.

Putting all these signs together the exponent of -1 changes by $|w_i| + |w_{i+1}| + (|w_i| + 1)(|w_{i+1}| + 1)$ on the left hand side of the equation and by $1+|w_i||w_{i+1}|$ on the right hand side, coinciding modulo 2. Thus the sign equation holds true for the transposition τ, which finishes the proof. □

Putting together the above discussion we get the following theorem:

Theorem 2.14. Let L be a $\mathbb{Z}$-graded vector space. Then there is a natural one-to-one correspondance between L_∞-algebra structures on L and differentials $D : S^\bullet(sL) \to S^\bullet(sL)$, where D is called a *differential* if and only if it is a degree-one coderivation satisfying $D^2 = 0$.

Proof. One direction follows directly from the Lemmata 2.6 and 2.13. For the other direction consider the map $\hat{l} := \pi_1 \circ D : S^\bullet(sL) \to sL$. The map $\hat{l}$ uniquely determines D since coderivations on $S^\bullet(sL)$ are uniquely determined by their values on the cogenerators[2]. Next the components $\hat{l}_i$ of $\hat{l}$ are modified in the following way: $l_i(x_1, ..., x_n) = (-1)^{\alpha(x_1,...,x_n)} s^{-1}\hat{l}_i(sx_1, ..., sx_n)$. Lemma 2.6 (read backwards) then guarantees that the constructed l_i are graded skew-symmetric. To prove equation (5) one has to see that (5) was not used to calculate (8). Then one considers (9) for $d = n$. Equation (5) follows now from the sign discussion after (9). □

Let us see, how classical objects fit in this setting:

Example 2.15. Let L be a differential graded Lie algebra. Then from the L_∞-perspective the only nonzero brackets are $l_1 : L \to L$ and $l_2 : L \times L \to L$. Thus the only nonzero brackets after the shift are $\hat{l}_2 : sL \times sL \to sL$ and $\hat{l}_1 : sL \to sL$ defined by:

$$\hat{l}_2(sx_1, sx_2) = (-1)^{\alpha(x_1,x_2)} sl_2(x_1, x_2) = (-1)^{|x_1|} sl_2(x_1, x_2),$$
$$\hat{l}_1(sx_1) = (-1)^{\alpha(x_1)} sl_1(x_1) = -sl_1(x_1).$$

[2]This is a direct concequence of corollary A.2 and the fact that $S^\bullet(sL)$ is a connected bicomodule over itself.

Thus we get the following formula for D:

$$D(sx_1 \odot ... \odot sx_n) = -\sum_{\sigma \in ush(1,n-1)} \epsilon(\sigma, sx_1, ..., sx_n) sl_1(x_{\sigma(1)}) \odot sx_{\sigma(2)} \odot ... \odot sx_{\sigma(n)}$$
$$+ \sum_{\sigma \in ush(2,n-2)} (-1)^{|x_1|} \epsilon(\sigma, sx_1, ..., sx_n) sl_2(x_{\sigma(1)}, x_{\sigma(2)}) \odot sx_{\sigma(3)} \odot ... \odot sx_{\sigma(n)}.$$

Evaluating D^2 on one, two and three elements would retrieve $l_1^2 = 0$, the graded Leibnitz identity and the graded Jacobi identity.

Let us further assume that $l_1 = 0$. This means that L is a graded Lie algebra (as a special case of a differential graded Lie algebra). Then the $\hat{l}_1$ term also vanishes and we get the following formula for D.

$$D(sx_1 \odot ... \odot sx_n) = \sum_{\sigma \in ush(2,n-2)} (-1)^{|x_1|} \epsilon(\sigma, sx_1, ..., sx_n) sl_2(x_{\sigma(1)}, x_{\sigma(2)}) \odot sx_{\sigma(3)} \odot ... \odot sx_{\sigma(n)}.$$

Next we regard a graded Lie algebra concentrated in even degree. Then all elements in sL are odd and thus all transpositions in any permutation are odd. Thus $\epsilon(\sigma, sx_1, ..., sx_n) = sgn(\sigma)$. This gives our formula for D the following form:

$$D(sx_1 \odot ... \odot sx_n) = \sum_{\sigma \in ush(2,n-2)} sgn(\sigma) sl_2(x_{\sigma(1)}, x_{\sigma(2)}) \odot sx_{\sigma(3)} \odot ... \odot sx_{\sigma(n)}.$$

We can rewrite the sum in terms of $i = \sigma(1)$ and $j = \sigma(2)$, where the $\hat{x}_k$ means that x_k is left out.

$$D(sx_1 \odot ... \odot sx_n) = \sum_{i \leq i < j \leq n} (-1)^{i+j-1} sl_2(x_i, x_j) \odot sx_1 \odot ... \odot s\hat{x}_i \odot ... \odot s\hat{x}_j \odot ... \odot sx_n.$$

As sL only has elements in odd degree $S^\bullet(sL)$ is the (ungraded) exterior algebra of sL denoted by $\Lambda^\bullet(sL)$. In this case we write $\wedge$ instead of $\odot$. Additionally we will restrict ourselves to the case, where $L = L_0 = \mathfrak{g}$ i.e. where L is an ordinary Lie algebra. Then regarding our D degree wise as a family of maps $D = D_k : \Lambda^k \mathfrak{g} \to \Lambda^{k-1}\mathfrak{g}$ and writing $[\cdot, \cdot]$ for l_2 we get:

$$D_k(x_1 \wedge ... \wedge x_k) = \sum_{i \leq i < j \leq k} (-1)^{i+j+1} [x_i, x_j] \wedge x_1 \wedge ... \wedge \hat{x}_i \wedge ... \wedge \hat{x}_j \wedge ... \wedge x_k.$$

We can regard $(\Lambda^\bullet \mathfrak{g}, D)$ as a differential graded vector space (with a degree -1 differential) as follows:

$$\Lambda^0\mathfrak{g} = \mathbb{K} \xleftarrow{\;0\;} \Lambda^1\mathfrak{g} = \mathfrak{g} \xleftarrow{\;D\;} \Lambda^2\mathfrak{g} \xleftarrow{\;D\;} \Lambda^3\mathfrak{g} \xleftarrow{\;D\;} \Lambda^4\mathfrak{g} \xleftarrow{\;D\;} \cdots$$

The above differential D is often written as ∂ in the literature but we will not use the letter ∂ for this differential here.

If $\mathfrak{g}$ is finite-dimensional we can regard the dual differential graded vector space $((\Lambda^\bullet \mathfrak{g})^*, D^*)$ $= ((\Lambda^\bullet \mathfrak{g}^*), -\delta)$, where $\delta = \delta^k : \Lambda^k \mathfrak{g}^* \to \Lambda^{k+1} \mathfrak{g}^*$ is the Chevalley-Eilenberg differential defined as follows:

$$(\delta^k(\phi))(x_1, ..., x_{k+1}) = \phi(-D_{k+1}(x_1 \wedge ... \wedge x_{k+1}))$$

$$= \phi\left(\sum_{1 \leq i < j \leq k+1} (-1)^{i+j}[x_i, x_j] \wedge x_1 \wedge ... \wedge \hat{x}_i \wedge ... \wedge \hat{x}_j \wedge ... \wedge x_{k+1} \right)$$

$$= \sum_{1 \leq i < j \leq k+1} (-1)^{i+j}\phi([x_i, x_j] \wedge x_1 \wedge ... \wedge \hat{x}_i \wedge ... \wedge \hat{x}_j \wedge ... \wedge x_{k+1}).$$

Thus, if L is a finite-dimensional Lie algebra, $D : S^\bullet(sL) \to S^\bullet(sL)$ is the dual of the *Chevalley-Eilenberg algebra* (also called *Chevalley-Eilenberg cochain complex*) of L.

2.3 Morphisms of L_∞-algebras

Motivated by the definition of morphisms of Lie algebras (linear maps satisfying $f([x, y]) = [f(x), f(y)]$) one arrives at the following definition of an L_∞-algebra morphism:

Definition 2.16. Let $(L, \{l_k \mid k \in \mathbb{N}\})$ and (L', l'_k) be L_∞ algebras . A *strict L_∞-morphism* from L to L' is a graded skew-symmetric linear map f of degree zero, satisfying the following condition for all $k \in \mathbb{N}$ and all $x_1, ..., x_k \in L$:

$$f(l_k(x_1, ..., x_k)) = l'_k(f(x_1), ..., f(x_k)) \tag{10}$$

For $k = 1$ this condition implies $f \circ l_1 = l'_1 \circ f$, which means that f is a morphism of differential graded vector spaces. The above defined morphisms are called strict, because (10) is a very restrictive condition, which only few maps will satisfy. What we often need is a weaker notion of morphism, which takes into account the relation of the different brackets. The description of L_∞-algebras provided by Theorem 2.14 makes is easier to define a weaker notion of morphism, as all relevant information is encoded into one map, instead of infinitely many.

Definition 2.17. Let $(L, \{l_k\})$ resp. $(L', \{l'_k\})$ be L_∞-algebras corresponding to $(S^\bullet(sL), D)$ resp. $(S^\bullet(sL'), D')$. A *(weak) L_∞-morphism* F from $(L, \{l_k\})$ to $(L', \{l'_k\})$ is a morphism of counital coalgebras $F : S^\bullet(sL) \to S^\bullet(sL')$ that preserves the differential. This means it is a degree-zero linear map, such that the following diagrams commute:

comultiplicativity:

$$\begin{array}{ccc} S^\bullet(sL) & \xrightarrow{F} & S^\bullet(sL') \\ \downarrow{\Delta} & & \downarrow{\Delta} \\ S^\bullet(sL) \otimes S^\bullet(sL) & \xrightarrow{F \otimes F} & S^\bullet(sL') \otimes S^\bullet(sL') \end{array}$$

counitality:

$$\begin{array}{ccc} S^\bullet(sL) & \xrightarrow{F} & S^\bullet(sL') \\ & {}_{\pi_0}\searrow & \downarrow{\pi_0} \\ & & \mathbb{K} \end{array}$$

differential-preservation

$$\begin{array}{ccc} S^\bullet(sL) & \xrightarrow{F} & S^\bullet(sL') \\ \downarrow{D} & & \downarrow{D'} \\ S^\bullet(sL) & \xrightarrow{F} & S^\bullet(sL') \end{array}$$

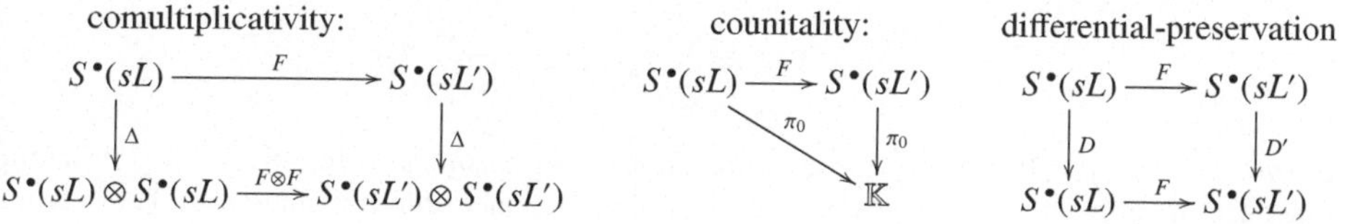

With the tools we have learned in the last section we will unpack this definition to formulate it in terms of multibracket identities. First of all we observe, that F is uniquely determined by its values on the cogenerators $sL' \subset S^\bullet(sL')$. This is again a consequence of the universal property of $S^\bullet(sL')$ (cf. Theorem A.1) and the fact that $S^\bullet(sL)$ is a connected cobimodule over $S^\bullet(sL')$. Thus we can regard F as a family $\{\hat{f}_k \mid k \geq 0\}$ of graded-symmetric linear maps $\hat{f}_k = \pi_1 \circ F|_{S^k(sL)} : S^k(sL) \to sL'$. Next we set

$$f_k(x_1, ..., x_k) = -(-1)^{\alpha(x_1,...,x_k)} s^{-1} \hat{f}_k(sx_1 \odot ... \odot sx_k),$$

where s^{-1} is the inverse of $s : L' \to sL'$. With a calculation analogous to the proof of Lemma 2.6 we can show that f_k are graded skew-symmetric maps of degrees $1-k$. The next step is to encode the preservation of the differential: $F \circ D = D' \circ F$. As $F \circ D$ is a coderivation from $S^\bullet(sL)$ to $S^\bullet(sL')$ it is uniquely determined by its values on the cogenerators by the universal property of $S^\bullet(sL')$.[3] Thus it is sufficient to reformulate the equation $\pi_1 \circ F \circ D = \pi_1 \circ D' \circ F$ in terms of the f_k. First let us rephrase the left hand side in terms of expressions involving $\hat{f}_k$, $\hat{l}_k$ and $\hat{l}'_k$.

$$\begin{aligned}
&\pi_1 \circ F \circ D(sx_1 \odot ... \odot sx_n) \\
&= \pi_1 \circ F\left(\sum_{i=1}^{n} \sum_{\sigma \in ush(i,n-i)} \epsilon(\sigma, sx_1, ..., sx_n)\left(\hat{l}_i(sx_{\sigma(1)} \odot ... \odot sx_{\sigma(i)})\right) \odot sx_{\sigma(i+1)} \odot ... \odot sx_{\sigma(n)}\right) \\
&= \sum_{i=1}^{n} \sum_{\sigma \in ush(i,n-i)} \epsilon(\sigma, sx_1, ..., sx_n)\hat{f}_{n-i+1}\left(\left(\hat{l}_i(sx_{\sigma(1)} \odot ... \odot sx_{\sigma(i)})\right) \odot sx_{\sigma(i+1)} \odot ... \odot sx_{\sigma(n)}\right) \\
&= \sum_{i+j=n+1} \sum_{\sigma \in ush(i,n-i)} \epsilon(\sigma, sx_1, ..., sx_n)\hat{f}_j\left(\left(\hat{l}_i(sx_{\sigma(1)} \odot ... \odot sx_{\sigma(i)})\right) \odot sx_{\sigma(i+1)} \odot ... \odot sx_{\sigma(n)}\right).
\end{aligned}$$

For the right hand side, we need a formula of F in terms of the $\hat{f}_k$. Equation (54) in [5] provides us with such a formula. With our conventions it takes the following form:

$$\begin{aligned}
F(sx_1 \odot ... \odot sx_n) &= \sum_{p=1}^{n} \sum_{\sum_{j=1}^{p} k_j = n} \sum_{\sigma \in ush(k_1,...,k_p)} \frac{\epsilon(\sigma, sx_1, ..., sx_n)}{p!} \times \\
&\hat{f}_{k_1}(sx_{\sigma(1)} \odot ... \odot sx_{\sigma(k_1)}) \odot \hat{f}_{k_2}(sx_{\sigma(k_1+1)} \odot ... \odot sx_{\sigma(k_1+k_2)}) \odot ... \odot \hat{f}_{k_p}(sx_{\sigma(n-k_p+1)} \odot ... \odot sx_{\sigma(n)}).
\end{aligned}$$

Here $\sigma \in ush(k_1, ..., k_p) \subset S_n$ are the permutations satisfying

$$\sigma(\sum_{i=1}^{j} k_i + 1) < \sigma(\sum_{i=1}^{j} k_i + 2)... < \sigma(\sum_{i=1}^{j+1} k_i),$$

for all $j \in 0, ..., p-1$. We call those permutations *multi-unshuffles*. Intuitively they unshuffle one deck of cards into p subdecks without distorting the inner ordering otherwise. They form a

[3] again according to Corollary A.2

representative system for $S_n/(S_{k_1} \times S_{k_2} \times ... \times S_{k_p})$. Next we apply $\pi_1 \circ D'$ to our formula for F and get:

$$\pi_1 \circ D \circ F(sx_1 \odot ... \odot sx_n) = \sum_{p=1}^{n} \sum_{\sum_{j=1}^{p} k_j = n} \sum_{\sigma \in ush(k_1,...,k_p)} \frac{\epsilon(\sigma, sx_1, ..., sx_n)}{p!} \times$$

$$\hat{l}'_p \left(\hat{f}_{k_1}(sx_{\sigma(1)} \odot ... \odot sx_{\sigma(k_1)}) \odot \hat{f}_{k_2}(sx_{\sigma(k_1+1)} \odot ... \odot sx_{\sigma(k_1+k_2)}) \odot ... \odot \hat{f}_{k_p}(sx_{\sigma(n-k_p+1)} \odot ... \odot sx_{\sigma(n)})\right).$$

We can get rid of the $\frac{1}{p!}$ by the cost of more complicated indices. To achieve that we have to somehow avoid the cases, where two summands are equal up to reordering. We can achieve that by demanding $k_1 \leq k_2 ... \leq k_p$ and $\sigma(\sum_{i=1}^{j-1} k_i + 1) < \sigma(\sum_{i=1}^{j} k_i + 1)$ whenever k_j and k_{j+1} are equal. Thus we get the following expression:

$$\pi_1 \circ D \circ F(sx_1 \odot ... \odot sx_n) = \sum_{p=1}^{n} \sum_{\substack{\sum_{j=1}^{p} k_j = n \\ k_i \leq k_{i+1}}} \sum_{\substack{\sigma \in ush(k_1,...,k_p) \\ \sigma(\sum_{i=1}^{j-1} k_i + 1) < \sigma(\sum_{i=1}^{j} k_i + 1) \\ \text{whenever } k_j = k_{j+1}}} \epsilon(\sigma, sx_1, ..., sx_n) \times$$

$$\hat{l}'_p \left(\hat{f}_{k_1}(sx_{\sigma(1)} \odot ... \odot sx_{\sigma(k_1)}) \odot \hat{f}_{k_2}(sx_{\sigma(k_1+1)} \odot ... \odot sx_{\sigma(k_1+k_2)}) \odot ... \odot \hat{f}_{k_p}(sx_{\sigma(n-k_p+1)} \odot ... \odot sx_{\sigma(n)})\right).$$

Thus the preservation of the differential is equivalent to:

$$\sum_{i+j=n+1} \sum_{\sigma \in ush(i,n-i)} \epsilon(\sigma, sx_1, ..., sx_n) \hat{f}_j \left(\left(\hat{l}_i(sx_{\sigma(1)} \odot ... \odot sx_{\sigma(i)})\right) \odot sx_{\sigma(i+1)} \odot ... \odot sx_{\sigma(n)}\right)$$

$$= \sum_{p=1}^{n} \sum_{\substack{\sum_{j=1}^{p} k_j = n \\ k_i \leq k_{i+1}}} \sum_{\substack{\sigma \in ush(k_1,...,k_p) \\ \sigma(\sum_{i=1}^{j-1} k_i + 1) < \sigma(\sum_{i=1}^{j} k_i + 1) \\ \text{whenever } k_j = k_{j+1}}} \epsilon(\sigma, sx_1, ..., sx_n) \hat{l}'_p \left(\hat{f}_{k_1}(sx_{\sigma(1)} \odot ... \odot sx_{\sigma(k_1)})\right.$$

$$\left. \odot \hat{f}_{k_2}(sx_{\sigma(k_1+1)} \odot ... \odot sx_{\sigma(k_1+k_2)}) \odot ... \odot \hat{f}_{k_p}(sx_{\sigma(n-k_p+1)} \odot ... \odot sx_{\sigma(n)})\right).$$

The next step is substituting $\hat{l}_i, \hat{f}_i$ with the unhatted versions. The following lemma is a slightly modified version of the last statement before Definition 4.7.1 in Section 4.7 of [1], where the necessary condition $k_i \leq k_{i+1}$ is not explicitly stated.

Lemma 2.18. A weak L_∞-morphism F is equivalently given by a family $\{f_k\}$ of graded skew-symmetric maps $f_k : \bigotimes^k L \to L$ of degrees $1-k$ satisfying the following condition for all $n \in \mathbb{N}$:

$$\sum_{i+j=n+1} \sum_{\sigma \in ush(i,n-i)} (-1)^{i(j+1)} sgn(\sigma) \epsilon(\sigma, x_1, ..., x_n) f_j \left((l_i(x_{\sigma(1)} \odot ... \odot x_{\sigma(i)})) \odot x_{\sigma(i+1)} \odot ... \odot x_{\sigma(n)}\right)$$

$$= \sum_{p=1}^{n} \sum_{\substack{\sum_{j=1}^{p} k_j = n \\ k_i \leq k_{i+1}}} \sum_{\substack{\sigma \in ush(k_1,...,k_p) \\ \sigma(\sum_{i=1}^{j-1} k_i + 1) < \sigma(\sum_{i=1}^{j} k_i + 1) \\ \text{whenever } k_j = k_{j+1}}} (-1)^{\beta} sgn(\sigma) \epsilon(\sigma, x_1, ..., x_n) \times$$

$$l'_p(f_{k_1}(x_{\sigma(1)} \odot ... \odot x_{\sigma(k_1)}) \odot f_{k_2}(x_{\sigma(k_1+1)} \odot ... \odot x_{\sigma(k_1+k_2)}) \odot ... \odot f_{k_p}(x_{\sigma(n-k_p+1)} \odot ... \odot x_{\sigma(n)})),$$

where β is given by the following formula:

$$\beta = \frac{p(p-1)}{2} + \sum_{i=1}^{p} k_i(p-i) +$$

$$(k_p - 1)\sum_{i=1}^{n-k_p} |x_{\sigma(i)}| + (k_{p-1} - 1)\sum_{i=1}^{n-(k_p+k_{p-1})} |x_{\sigma(i)}| + ... + (k_2 - 1)\sum_{i=1}^{n-(k_p+k_{p-1}+...+k_2)} |x_{\sigma(i)}|.$$

Proof. Again, to prove the lemma, we just have to show, that the signs are well-behaved. For the left-hand-side of the equation we make use of the calculation from Lemma 2.13, plus one additional step: in 2.13 we were only interested in the relative sign and did not care about the absolute one. Here we need the absolute one, too. For that, let us just calculate the sign difference between $f_j(l_i(x_1, ..., x_i), x_{i+1} ..., x_{i+j-1})$ and $\hat{f}_j(\hat{l}_i(sx_1, ..., sx_i), sx_{i+1} ..., sx_{i+j-1})$. By definition, the sign is $-(-1)^{\alpha(x_1,...,x_i)} \cdot (-1)^{\alpha(l_i(x_1,...,x_i), x_{i+1},...,x_{i+j-1})}$. With the case distinction from 2.13 we see that $\alpha(x_1, ..., x_i) + \alpha(l_i(x_1, ..., x_i), x_{i+1}, ..., x_{i+j-1}) \equiv 1 + \alpha(x_1, ..., x_{i+j-1}) \mod 2$. Thus, the degree shift on the left hand side of the equation yields an absolute sign of $-(-1)(-1)^{\alpha(x_1,...,x_n)} = (-1)^{\alpha(x_1,...,x_n)}$.

For the right hand side of the equation we will proceed as follows:

1. We will show, that the signs agree for $\sigma = id$ and $|x_i| \equiv 0 \mod 2 \ \forall i$.
2. We will show, that they stay correct for arbitrary x_1.
3. We will finish by showing, that they stay correct for transpositions and thus for arbitrary permutations.

Step 1: Let $\sigma = id$ and $|x_i| \equiv 0 \mod 2 \ \forall i$, then $(-1)^{\alpha(x_1,...,x_n)} = (-1)^n$. Let us define, by a slight abuse of notation, $\alpha_p : \mathbb{Z}^p \to \mathbb{Z}$ by

$$\alpha_p(l_1, ..., l_p) = \sum_{\substack{i \text{ is odd} \\ i \in \{1,...,p\}}} l_i \quad \text{when p is even,}$$

$$\alpha_p(l_1, ..., l_p) = 1 + \sum_{\substack{i \text{ is even} \\ i \in \{1,...,p\}}} l_i \quad \text{when p is odd.}$$

Then we have $\alpha(x_1, ..., x_p) = \alpha_p(|x_1|, ..., |x_p|)$. Furthermore, on the right hand side we have the sign from applying l'_p to $f_{k_1}(x_1, ..., x_{k_1}), ..., f_{k_p}(x_{n-k_p+1}, ..., x_n)$. For y_i of even degree, the degree of $f_{k_i}(y_1, ..., y_{k_i})$ is $1 - k_i \mod 2$. Thus the sign from applying l' to the f's is $(-1)^{\alpha_p(1-k_1,...,1-k_p)}$. The sign of applying f_{k_i} to elements of even degree is $(-1)^{1+\alpha_{k_i}(0,...,0)} = (-1)^{1+k_i}$. Thus in total we get the sign $(-1)^c$, where

$$c = \alpha(1 - k_1, ..., 1 - k_p) + \sum_{i=1}^{p}(1 + k_i).$$

Upon consideration of the case where p is even and where p is odd separately, one can verify the following alternative formula for α_p:

$$\alpha(l_1, ..., l_p) \equiv p + \sum_{i=1}^{p} l_i(p-i) \mod 2 \tag{11}$$

From (11) we deduce:

$$\begin{aligned}
c &\equiv p + \sum_{i=1}^{p}(1-k_i)(p-i) + \sum_{i=1}^{p}(1+k_i) \equiv p + \sum_{i=1}^{p} k_i(p-i) - \sum_{i=1}^{p}(p-i) + \sum_{i=1}^{p}(1+k_i) \\
&\equiv p + \sum_{i=1}^{p} k_i(p-i) - p^2 + \sum_{i=1}^{p} i + \sum_{i=1}^{p}(1+k_i) \equiv p + \sum_{i=1}^{p} k_i(p-i) - p^2 + \sum_{i=1}^{p} i + p + n \\
&\equiv p + \sum_{i=1}^{p} k_i(p-i) - p^2 + \frac{p(p+1)}{2} + p + n \equiv (p-p^2) + \sum_{i=1}^{p} k_i(p-i) + \left(p + \frac{p(p+1)}{2}\right) + n \\
&\equiv \sum_{i=1}^{p} k_i(p-i) + \frac{p(p-1)}{2} + n \mod 2.
\end{aligned}$$

As the degrees of all elements are even, $\beta \equiv \sum_{i=1}^{p} k_i(p-i) + \frac{p(p-1)}{2} \mod 2$. So the sign we get on the right hand side of the equation is precisely $(-1)^{\beta}(-1)^n$ and the sign we get on the left hand side is $(-1)^n$. Thus the signs are well-behaved in the case of the identity permutation and elements of even degree.

Step 2: Next we show, that the sign determined by β stays correct, when we change the degrees of our elements. As k_i, p and n do not change, the relevant part of β is:

$$(k_p - 1)\sum_{i=1}^{n-k_p} |x_i| + (k_{p-1} - 1)\sum_{i=1}^{n-(k_p+k_{p-1})} |x_i| + ... + (k_2 - 1)\sum_{i=1}^{n-(k_p+k_{p-1}+...+k_2)} |x_i|$$

So assume we change[4] the degree of x_t by one for a certain $t \in \{1, ..., n\}$. There exists some $r \in \{0, 1, ..., p-1\}$ such that $\sum_{j=1}^{r} k_j < t \leq \sum_{j=1}^{r+1} k_j$. Thus $t \in I_s$ for all $s \in \{0, 1, ..., p-(r+1)\}$, where the index sets I_s are defined by $I_s = \{1, ..., n-(k_p + k_{p-1} + ... + k_{p-s})\}$. If we increase the degree of x_t by one, β changes by $(k_p - 1) + ... + (k_{r+2} - 1) = n - (k_1 + ... + k_{r+1}) + (p - r - 1) =: \Delta\beta$. When we look at the change from the $\alpha(...)$ perspective, we notice that on the right hand side of the equation two alphas can change: The $\alpha = \alpha_p$ which belongs to l'_p and $\alpha_{k_{r+1}}$ which belongs to $f_{k_{r+1}}$. Using formula (11) we see that α_p changes by $p-(r+1) \mod 2$, when the degree of $f_{k_{r+1}}(x_{k_1+...+k_r+1}, ..., x_{k_1+...+k_{r+1}})$ increases by one. Analogously we see that:

$$\alpha_{k_{r+1}}(x_{k_1+...+k_r+1}, ..., x_{k_1+...+k_{r+1}}) \equiv k_{r+1} + \sum_{m=1}^{k_{r+1}} x_{(k_1+...+k_r+m)}(k_{r+1} - m) \mod 2.$$

[4]we do not care, whether we increase or decrease as both changes behave the same with respect to signs

If we increase the degree of x_t by one, α_{k_r} changes by $k_{r+1} - m \equiv k_{r+1} - (t - (k_1 + ... + k_r)) \mod 2$. Thus, mod 2, we get the followng change in the exponent of the sign on the right hand side of the equation:

$$(p - r - 1) + k_{r+1} - (t - (k_1 + ... + k_r)) \equiv (p - (r + 1)) + k_1 + ... + k_{r+1} - t \mod 2$$

On the left hand side of the equation we had the sign belonging to $\alpha(x_1, ..., x_n)$. By formula (11) the change from increasing the degree of x_t by one would be $n - t \mod 2$. So all in all increasing the degree of some x_t by one gives us the sign change $(-1)^c$, where

$$\begin{aligned} c \equiv (p - r - 1) + k_1 + ... + k_{r+1} - t + (n - t) &\equiv (p - r - 1) + k_1 + ... + k_{r+1} + n \\ &\equiv (p - r - 1) + (n - (k_1 + ... + k_{r+1})) \equiv \Delta\beta \mod 2 \end{aligned}$$

So the sign change resulting from increasing the degree of x_t resembles $(-1)^{\Delta\beta}$.

Step 3: Now it remains to show, that the signs behave in the right way, when we apply transpositions. The calculation for the left hand side of the equation is analogous to the corresponding part of the proof of 2.13. For the right hand side, the only transpositions of interest to us are the ones interchanging elements from different blocks. As it is enough to regard transpositions of neighbouring elements we can assume, without loss of generality, that τ interchanges x_m and x_{m+1}, where $m = k_1 + ... + k_r$. Let us first look at the sign changes relevant to our formula:

- $sgn(\tau) = -1$ as τ is a transposition
- $\epsilon(\tau, x_1, ..., x_n) = (-1)^{|x_m||x_{m+1}|}$
- β changes by $\Delta\beta = (|x_{m+1}| - |x_m|) \cdot (k_{r+1} - 1)$. This is because there is only one summand, which includes x_m and does not include x_{m+1} in β. It is the summand $(k_{r+1} - 1) \sum_{i=1}^{n-(k_p+...+k_{r+1})} |x_i|$. By interchanging x_m and x_{m+1} this summand changes by $(|x_{m+1}| - |x_m|) \cdot (k_{r+1} - 1)$.

So we have an overall sign change of $(-1)^{1+|x_m||x_{m+1}|+(|x_{m+1}|-|x_m|)\cdot(k_{r+1}-1)}$. Let us now see, what sign we should get according to our definition of f_i:

- $\epsilon(\tau, sx_1, ..., sx_n) = (-1)^{(1+|x_m|)(1+|x_{m+1}|)}$
- α_p, the sign belonging to $\hat{l}_p \to l_p$ changes by $\Delta\alpha_p \equiv |x_m| - |x_{m+1}| \mod 2$. To see that, observe, that the transposition changes the degree of $f_{k_r}(x_{k_1+...+k_{r-1}+1}, ..., x_m)$ by $|x_{m+1}| - |x_m|$ and the degree of $f_{k_{r+1}}(x_{m+1}, ..., x_{m+k_{r+1}})$ by $|x_m| - |x_{m+1}|$. As $\alpha_p(f_{k_1}(x_1, ..., x_{k_1}), ..., f_{k_p}(x_{n-k_p+1}, ..., x_n))$ takes into account exactly one of those two, the assertion then follows from the congruence $|x_m| - |x_{m+1}| \equiv |x_{m+1}| - |x_m| \mod 2$.
- α_{k_r}, the sign belonging to $f_{k_r}(x_{k_1+...+k_{r-1}+1}, ..., x_{k_m})$ does not change, as α never depends on the degree of the last element.
- $\alpha_{k_{r+1}}$ changes only if it depends on the first element. This is the case if k_{r+1} is even. The value then changes by $|x_m| - |x_{m+1}|$. This means $\Delta\alpha_{k_{r+1}} \equiv (k_{r+1} - 1)(|x_m| - |x_{m+1}|) \mod 2$.

So we get an overall sign change of $(-1)^{(1+|x_m|)(1+|x_{m+1}|)+|x_m|-|x_{m+1}|+(k_{r+1}-1)(|x_m|-|x_{m+1}|)}$, which corresponds to the sign change according to our formula and thus finishes the proof. □

With this lemma we now have a description of weak L_∞-morphisms in terms of multi-bracket identities. Restricting ourselves to the case, where $f_k = 0$ for all $k > 1$ gives us back the definition of strict morphisms:

Example 2.19. A strict L_∞-morphism is a weak L_∞-morphism satisfying $f_k = 0$ for all $k > 1$.

Another example of interest is the case, where the target L_∞-algebra is a differential graded Lie algebra (i.e. $l'_k = 0$ for $k > 2$). In accordance with [6] and [1] we get in this case the following outcome:

Example 2.20. An L_∞-morphism $(L, l_k) \to (L', l'_k)$, with target $(L', l'_k) = (L', d, [\cdot,\cdot])$ a differential graded Lie algebra, is a family of maps $\{f_k\}$ of degrees $1-k$ satisfying the following conditions:

$$\sum_{i+j=n+1} \sum_{\sigma \in ush(i,n-i)} (-1)^{i(j+1)} sgn(\sigma)\epsilon(\sigma, x_1, ..., x_n) f_j\left((l_i(sx_{\sigma(1)} \odot ... \odot sx_{\sigma(i)})) \odot sx_{\sigma(i+1)} \odot ... \odot sx_{\sigma(n)}\right)$$

$$= df_n(x_1, ..., x_n) + \sum_{s+t=n} \sum_{\substack{\sigma \in ush(s,t) \\ \sigma(1)<\sigma(s+1)}} sgn(\sigma)\epsilon(\sigma, x_1, ..., x_n)(-1)^{1+s+(t-1)\sum_{i=1}^{s}|x_{\sigma(i)}|}[f_s(x_1, ..., x_s), f_t(x_{s+1}, .., x_n)].$$

In this description we have set $s = i_1$ and $t = i_2$. Instead of demanding $s \leq t$ and $\sigma(1) < \sigma(s+1)$ in case of equality, we just demand $\sigma(1) < \sigma(s+1)$ in all cases. This leads to the same summands as before, the elements we lose by demanding $\sigma(1) < \sigma(s+1)$ for $s < t$ are exactly the elements we get by allowing $s > t$.

2.4 Representations of L_∞-algebras

Before we define representations of L_∞-algebras let us review the case of Lie algebras.

Definition 2.21. Let $\mathfrak{g}$ be a Lie algebra and V be a vector space. A *representation of* $\mathfrak{g}$ *on* V is a Lie algebra morphism $\rho : \mathfrak{g} \to End(V)$, where $End(V)$, the space of linear endomorphisms of V, carries the structure of a Lie algebra defined by the commutator bracket $[A, B] = AB - BA$.

Accordingly it seems natural to define a representation of an L_∞-algebra as an L_∞-morphism into the endomorphism space of some appropriate representation space. As an example we look at the following type of Lie algebra representations:

Example 2.22. Let $\mathfrak{g}$ be a Lie algebra and $End(\mathfrak{g})$ the space of linear endomorphisms of its underlying vector space. Then $ad : \mathfrak{g} \to End(\mathfrak{g})$ defined by $ad_x(\cdot) = [x, \cdot]$ is a representation of $\mathfrak{g}$, called *the adjoint representation.*

If we want the representation theory of L_∞-algebras to be a generalization of the theory for Lie algebras, the representation space should carry some of the structure underlying an L_∞-algebra. The most basic object underlying an L_∞-algebra in this setting is the graded vector space $L = \bigoplus L_i$. This object, however, does not relate the different degrees. In order to relate them, we need the differential $d = l_1$ which is also part of the information an L_∞-algebra provides. Thus our representation spaces should be differential graded vector spaces.

Definition 2.23. Let $V = \bigoplus_{i \in \mathbb{Z}} V_i$ be a differential graded vector space with differential d (of degree 1). We define $(End(V), \delta, [\cdot, \cdot])$ as follows:

- $End(V) = \bigoplus_{i \in \mathbb{Z}} End_i(V)$ as a graded vector space, where $End_i(V)$ are linear maps from V to V of degree i
- $\delta : End_i(V) \to End_{i+1}(V)$ is defined by $\delta(f) := d \circ f - (-1)^{|f|} f \circ d$
- $[\cdot, \cdot] : End_i(V) \times End_j(V) \to End_{i+j}(V)$ is defined by $[f, g] := f \circ g - (-1)^{|f||g|} g \circ f$

It is easy to check that $(End(V), \delta, [\cdot, \cdot])$ is a differential graded Lie algebra.

Luckily, a differential graded Lie algebra is a special case of an L_∞-algebra, so we can use the above construction to define representations of L_∞-algebras.

Definition 2.24. Let L be an L_∞-algebra and V a differential graded vector space. A *representation of L on V* is an L_∞-morphism from L to $End(V)$ i.e. a family of graded skew-symmetric maps f_k satisfying the conditions explicited in Example 2.20.

This generalization is reasonable, because if L is an ordinary Lie algebra and V a vector space (i.e. a differential graded vector space concentrated in degree zero with differential equal to zero) this definition reduces to Definition 2.21.

Unfortunately the above definition is not sufficient when one works in the setting of topological vector spaces and topological algebras. Demanding the map $L \to End(V)$ to be continuous is too restrictive even for for ordinary Lie algebras. Instead of considering the representation maps $L \to Hom(V, V)$ one regards their "adjoint morphisms" from $L \times V \to V$. This leads to the following definition:

Definition 2.25. Let $(L, [\cdot, \cdot])$ be a Lie algebra and V a vector space. An *L-module structure on V* is a bilinear map $m : L \times V \to V$ satisfying $m([x, y], v) = m(x, (m(y, v))) - m(y, m(x, v))$.

For a representation $\rho : L \to End(V)$ one can define a module structure on V by $m(x, v) := \rho(x)(v)$ (The requirement needed for m follows from the Jacobi identity). Indeed the converse is also true. The above contruction defines a one-to-one correspondance between L-module structures on V and representations of L on V. This fact can be used to define a sensible notion of continuity on representations. One simply requires that the corresponding module map is continuous instead of requiring the representation map to be continuous itself.

Example 2.26. The map $m := [\cdot, \cdot] : L \times L \to L$ turns L into a Lie module over itself. This is exactly the module structure belonging to the adjoint representation.

Generalizing Definition 2.25 yields the following definition of an L_∞-module:

Definition 2.27. Let $L = (L, \{l_k\})$ be an L_∞-algebra and (V, d) a differential graded vector space. An *L-module structure on V* is a family of graded skew-symmetric multilinear maps

$\left\{m_k : \left(\times^{k-1} L\right) \times V \to V \,\middle|\, k \in \mathbb{N}\right\}$ of degrees 2-k satisfying $m_1 = d$ and the following condition for all $x_1, ..., x_{n-1} \in L$ and $x_n \in M$:

$$0 = \sum_{i+j=n+1} (-1)^{i(j+1)} \sum_{\substack{\sigma \in ush(j,n-j) \\ \sigma(n)=n}} sgn(\sigma)\epsilon(\sigma, x_1, ..., x_n) m_i(l_j(x_{\sigma(1)}, ..., x_{\sigma(j)}), x_{\sigma(j+1)}...x_{\sigma(n)})$$

$$+ \sum_{i+j=n+1} (-1)^{i(j+1)} \sum_{\substack{\sigma \in ush(j,n-j) \\ \sigma(j)=n}} sgn(\sigma)\epsilon(\sigma, x_1, ..., x_n)(-1)^{\beta} m_i(x_{\sigma(j+1)}...x_{\sigma(n)}, m_j(x_{\sigma(1)}, ..., x_{\sigma(j)})),$$

where $\beta = (j-1) + (i + |x_{\sigma(1)}| + ... + |x_{\sigma(j)}|)(|x_{\sigma(j+1)}| + ... + |x_{\sigma(n)}|)$ and ϵ is here defined with respect to the graded vector space $L \oplus V$. We denote the set of those modules by $mod_L(V)$.

Remark 2.28. If $L = V$ and $m_k = l_k$ the condition reduces to Equation (5). In the general setting we have to distinguish between unshuffles, which leave x_n in the position, which makes m_i applicable, and unshuffles moving x_n to the j-th position, so that m_j is applicable as the inner bracket. In order to apply m_i, we then have to move $m_j(x_{\sigma(1)}, ..., x_{\sigma(j)})$ to the last position, which gives us the additional sign of $(-1)^{\beta}$.

As in the Lie algebra case modules are an equivalent description of representations.

Theorem 2.29. Let L be an L_∞-algebra and V a differential graded vector space. Then the following map is a bijection:

$$mod_L(V) \to Hom_{L_\infty}(L, End(V))$$
$$\{m_k\}_{k\in\mathbb{N}} \mapsto \{f_k\}_{k\in\mathbb{N}} \quad \text{where } f_k(x_1, ..., x_k)(v) := (-1)^k m_{k+1}(x_1, ..., x_k, v).$$

A proof of this theorem can be found after Theorem 5.4 in [6].

Using this result, we can now define the adjoint representation of an L_∞-algebra in terms of L_∞-modules:

Example 2.30. Let L be an L_∞-algebra. Setting $m_k := l_k$ turns L into an L_∞-module over itself and thus, by Theorem 2.29, gives us a representation of L on its underlying differential graded vector space. This representation is called the *adjoint representation of L* as it forms a generalization of the adjoint representation of Lie algebras.

2.5 Lie m-algebras

A general L_∞-algebra is an infinite deformation of a Lie algebra consisting of infinitely many multi-brackets and infinitely many identities generalizing the Jacobi identity. For our applications a finite deformation is often sufficient. These "finite" deformations are called Lie m-algebras.

Definition 2.31. Let $m \in \mathbb{N}$. An L_∞-algebra L is called *Lie m-algebra* if its underlying graded vector space is concentrated in the degrees $\{k \in \mathbb{Z} \mid -(m-1) \leq k \leq 0\}$.

Restricting the allowed grading for the underlying vector space also imposes some conditions on the multi-bracket structure of a Lie m-algebra:

- $l_k = 0$ for $k \geq m + 2$, because l_k has degree $2 - k$.
- Equation (5) is only interesting for $n < m + 3$, because $l_i(l_j(...), ...)$ has degree $2 - i + 2 - j = 4 - (n + 1) = 3 - n$.

For the naming to make sense a Lie 1-algebra should be an ordinary Lie algebra. With the above observations we get:

Example 2.32. For $m = 1$ we get back the definition of an ordinary Lie algebra. The map l_k is non-zero only when $k = 2$ and Equation (5) is only interesting for n+1=3 and equals, in fact, the Jacobi identity.

Lie m-algebras are very "small" from the L_∞-perspective: Not only is the vector space concentrated in finitely many degrees, also there are only finitely many brackets and finitely many equations they have to satisfy. Nethertheless the conditions morphisms have to fulfil are rather complicated. There is, however, a class of Lie m-algebras, which is much easier to handle, but still includes many important examples. This class of "grounded" Lie m-algebras allows non-trivial brackets only in the "ground" degree zero.

Definition 2.33. A Lie m-algebra L is called *grounded*, if $l_k(x_1, ..., x_n)$ is zero whenever $k > 1$ and $\sum_{i=1}^{n} |x_i| \neq 0$.

Remark 2.34. Grounded Lie m-algebras are called "Lie m-algebras satisfying Property P" in [5], cf. Subsection 3.2 therein.

Grounded Lie m-algebras are much simpler than general ones, as higher brackets are non-trivial only for ground degree elements. This also simplifies Equation (5). Let us take a look at the nonvanishing summands. In any case the result from applying the inner bracket l_i has degree less or equal to $2-i$, except in the case where $j = 1$. Thus Equation (5) for a given $2 \leq n \leq m + 2$ takes the following form:

$$(-1)^{n+1} l_1(l_n(x_{\sigma(1)} ..., x_{\sigma(n)})) + \sum_{\sigma \in ush(1,n-1)} sgn(\sigma)\epsilon(\sigma, x_1, ..., x_n) l_n(l_1(x_{\sigma(1)}), x_{\sigma(2)}, ..., x_{\sigma(n)})$$
$$+(-1)^{n-1} \sum_{\sigma \in ush(2,n-2)} sgn(\sigma)\epsilon(\sigma, x_1, ..., x_n) l_{n-1}(l_2(x_{\sigma(1)}, x_{\sigma(2)}), x_{\sigma(3)}, ..., x_{\sigma(n)}) = 0.$$

The first and the last of the three summands are nonzero, only if $|x_i| = 0$ for all i. The second summand is nonzero only if $|x_k| = -1$ for some k and $|x_i| = 0$ otherwise. Thus for $\sum |x_i| = 0$ the equation takes the following form, where the Koszul sign ϵ can be left out, as all elements have degree zero

$$l_1(l_n(x_1, ..., x_n)) + \sum_{\sigma \in ush(2,n-2)} sgn(\sigma) l_{n-1}(l_2(x_{\sigma(1)}, x_{\sigma(2)}), x_{\sigma(3)}, ..., x_{\sigma(n)}) = 0,$$

that (upon using that σ just pulls some x_i and x_j to the front) can also be written as follows:

$$l_1(l_n(x_1..., x_n)) + \sum_{1\leq i<j\leq n} (-1)^{i+j+1} l_{n-1}(l_2(x_i, x_j), x_1, ..., \hat{x}_i, ..., \hat{x}_j, ..., x_n) \quad = \quad 0 \qquad (12)$$

For $\sum |x_i| = -1$ we can leave out the permutations and signs as only the permutation pulling x_k with $|x_k| = -1$ to the front can give us a nonzero summand and we get:

$$l_n(l_1(x_k), x_1, ..., \hat{x}_k, ..., x_n) = 0 \qquad (13)$$

Therefore a grounded Lie m-algebras can be characterized as follows:

Lemma 2.35. A grounded Lie m-algebra can equivalently be described as a differential graded vector space $L = \bigoplus_{i=-m+1}^{0} L_i$ with a differential l_1 and a family of skew-symmetric multilinear maps $l_i : \bigtimes^i L_0 \to L$ of degrees $2 - i$ for $i \in \{2, ..., m + 1\}$ which satisfy the equations (12) and (13) for $1 \leq n \leq m + 1$.

Of course, the simpler structure of grounded Lie m-algebras also has an effect to their morphisms. Later on we will be especially interested in morphisms from Lie algebras into grounded Lie m-algebras, so we are inclined to present explicit formulas for such morphisms.

Lemma 2.36. Let $(\mathfrak{g}, [\cdot, \cdot])$ be a Lie algebra and $(L, \{l_1, ..., l_{m+1}\})$ be a grounded Lie m-algebra. An L_∞-morphism from $\mathfrak{g}$ to L is a collection of skew-symmetric multilinear maps $\{f_k \mid 1 \leq k \leq m\}$, where $f_k : \bigtimes^k \mathfrak{g} \to L$ has degree $1 - k$, such that the following equation holds for $2 \leq n \leq m + 1$:

$$\sum_{\sigma\in ush(2,n-2)} sgn(\sigma) f_{n-1}([x_{\sigma(1)}, x_{\sigma(2)}], x_{\sigma(3)}, ..., x_{\sigma(n)}) = l_1(f_n(x_1, ..., x_n)) + l_n(f_1(x_1), ..., f_1(x_n)),$$

for all $x_1, ..., x_n \in \mathfrak{g}$, and where f_{m+1} is to be interpreted as the zero map.

Proof. Follows directly from 2.18 by leaving out all zero summands and all degree-dependant signs (as the Lie algebra has no odd-degree elements). □

Remark 2.37. As σ just pulls some two elements x_i and x_j to the front its sign is $(-1)^{i+j}$. Thus we can rewrite the formula in Lemma 2.36 as follows:

$$\sum_{1\leq i<j\leq n} (-1)^{i+j+1} f_{n-1}([x_i, x_j], x_1, ..., \hat{x}_i, ..., \hat{x}_j, ..., x_n) = l_1(f_n(x_1, ..., x_n)) + l_n(f_1(x_1), ..., f_1(x_n)),$$

for all $x_1, ..., x_n \in \mathfrak{g}$ and where the hats over x_i and x_j denote that those elements are missing.

An important example of grounded Lie m-algebras are Lie algebras equiped with cocycles from Lie algebra cohomology. In the case of $m = 2$ every (finite-dimensional) Lie 2-algebra is equivalent to a Lie 2-algebra arising this way.[5] To illustrate this example let us first review the definition of Lie algebra cohomology with values in a representation space.

[5] See [2] for a detailed discussion and a proof.

Definition 2.38. Let $\mathfrak{g}$ be a Lie algebra, V a vector space and $\rho : \mathfrak{g} \to End(V)$ a representation. We define the *Lie algebra cohomology of* $\mathfrak{g}$ *with values in* V as the cohomology of the following differential graded vector space (cochain complex)

$$\Lambda^0(\mathfrak{g}, V) \xrightarrow{\delta} \Lambda^1(\mathfrak{g}, V) \xrightarrow{\delta} \Lambda^2(\mathfrak{g}, V) \xrightarrow{\delta} \Lambda^3(\mathfrak{g}, V) \xrightarrow{\delta} \Lambda^4(\mathfrak{g}, V) \xrightarrow{\delta} \dots \quad ,$$

where $\Lambda^k(\mathfrak{g}, V)$ is the vector space of skew-symmetric multilinear maps from $\bigtimes^k \mathfrak{g}$ to V and δ is defined as follows:

$$(\delta f)(x_1, ..., x_{n+1}) := \sum_{i=1}^{n+1}(-1)^{i+1}\rho(x_i)\,(f(x_1, ..., \hat{x}_i....x_n)) + \sum_{1\leq i<j\leq n+1} (-1)^{i+j} f([x_i, x_j], x_1, ..., \hat{x}_i, ..., \hat{x}_j, .., x_{n+1})).$$

Remark 2.39. One easily checks that $\delta^2 = 0$. In the case where $V = \mathbb{K}$ and $\rho = 0$ we get (up to sign) the Chevalley-Eilenberg complex from Example 2.15.

With this construction we can now exhibit a useful class of grounded Lie m-algebras.

Example 2.40. Let $\rho : \mathfrak{g} \to End(V)$ be a representation of a Lie algebra and $\eta \in \Lambda^{m+1}(\mathfrak{g}, V)$ a closed element with $m > 1$. Then we can define the following grounded Lie m-algebra:

- $L_0 := \mathfrak{g}$, $L_{-(m-1)} := V$, and $L_i := 0$ for all other i.
- $l_2 := [\cdot, \cdot]$ on $L_0 \times L_0$ and zero otherwise, $l_{m+1} := \eta$ on $\bigtimes^{m+1} L_0$ and zero otherwise. All other l_i are identically zero.

As l_1 is zero, Equation (13) holds, and the first summand of (12) vanishes, so that Condition (12) is equivalent to $\delta\eta = 0$, which is exactly the condition for η to be a cocycle. Thus the construction above yields indeed a grounded Lie m-algebra.

Another important example of Lie m-algebras (for $m = n$) are the observables of (pre-)n-plectic manifolds which will form the subject of the next chapter.

3 Observables of pre-n-plectic manifolds

Throughout this and the next chapter we fix $\mathbb{R}$ as our ground field.

3.1 Generalizing symplectic manifolds

In this section we define n-plectic manifolds as a generalization of (pre-)symplectic manifolds. Let us first review the notion of symplectic manifolds:

Definition 3.1. A *pre-symplectic manifold* (M, ω) is a manifold M equiped with a closed 2-form $\omega \in \Omega^2_{cl}(M)$. It is called *symplectic* if ω is non-degenerate, i.e., at any given point $p \in M$ the map $v \mapsto \iota_v\omega$ is an ismorphism from T_pM to T^*_pM.

The pointwise isomorphisms extend to an isomorphism of sections, i.e., we get a $C^\infty(M)$-module isomorphism:

$$\omega^\flat : \mathfrak{X}(M) = \Gamma(M, TM) \to \Gamma(M, T^*M) = \Omega^1(M), \quad v \mapsto \iota_v\omega.$$

On a symplectic manifold (M, ω), one can endow $C^\infty(M)$ with the structure of a Lie algebra. Given two smooth functions f, g one regards $df, dg \in \Omega^1(M)$ and transports them back with $(\omega^\flat)^{-1}$ to get $v_f := -(\omega^\flat)^{-1}(df)$ and $v_g := -(\omega^\flat)^{-1}(dg)$. The Poisson-bracket $(\{\cdot,\cdot\})$ of f with g is then defined by $\{f, g\} := \omega(v_f, v_g)$. The skew-symmetry follows directly from the skew-symmetry of ω and the Jacobi identity follows from the closedness of ω via direct calculation, which will be omitted here, as we will prove a more general statement later. This construction behaves well in the sense, that it turns $-(\omega^\flat)^{-1} \circ d : C^\infty(M) \to \mathfrak{X}(M)$ into a Lie algebra homomorphism. We will denote the image of this homomorphism by $\mathfrak{X}_{Ham}(M, \omega)$ and call the elements in $\mathfrak{X}_{Ham}(M, \omega)$ *Hamiltonian vector fields.*

In the pre-symplectic case $\omega^\flat$ might not be invertible. To fix this problem one defines a Lie algebra structure on $\widetilde{C^\infty}(M, \omega) := \{(v, f) \in \mathfrak{X}(M) \times C^\infty(M) \mid \omega^\flat(v) = -df\}$ instead of $C^\infty(M)$. This new object is the fiber product of $C^\infty(M)$ and $\mathfrak{X}(M)$, which means it fits into the following diagram and satisfies a universal property in its position there:

$$\begin{array}{ccc} \widetilde{C^\infty}(M,\omega) & \longrightarrow & \mathfrak{X}(M) \\ \downarrow & & \downarrow {\scriptstyle -\omega^\flat} \\ C^\infty(M) & \xrightarrow{\ d\ } & \Omega^1(M) \end{array} \tag{14}$$

The Lie bracket on $\widetilde{C^\infty}(M, \omega)$ is defined by $\{(v, f), (w, g)\} := ([v, w], \omega(v, w))$. If M is symplectic, $\omega^\flat$ is an isomorphism, and, implying that $\widetilde{C^\infty}(M, \omega)$ and $C^\infty(M)$ are isomorphic as Lie algebras.

In the general case let us first regard pre-n-plectic manifolds and treat n-plectic manifolds as a special case later on.

Definition 3.2. A *pre-n-plectic* manifold (M, ω) is a manifold M equipped with a closed $(n+1)$-form $\omega \in \Omega^{n+1}_{cl}(M)$.

The naming is shifted by one degree so that pre-symplectic manifolds are pre-1-plectic manifolds. As $\omega^\flat : \mathfrak{X}(M) \to \Omega^n(M)$, $v \mapsto \iota_v\omega$ yields now an element in $\Omega^n(M)$ the fiber product diagram (14) takes the following more general form here:

$$\begin{array}{ccc} \widetilde{\Omega^{n-1}}(M,\omega) & \xrightarrow{\pi_{\mathfrak{X}}} & \mathfrak{X}(M) \\ \downarrow{\scriptstyle \pi_\Omega} & & \downarrow{\scriptstyle -\omega^\flat} \\ \Omega^{n-1}(M) & \xrightarrow{d} & \Omega^n(M) \end{array} \tag{15}$$

where $\widetilde{\Omega^{n-1}}(M, \omega)$ is defined by $\widetilde{\Omega^{n-1}}(M, \omega) := \{(v, \alpha) \in \mathfrak{X}(M) \times \Omega^{n-1}(M) \mid \omega^\flat(v) = -d\alpha\}$. Now we would like to define a grounded Lie n-algebra based on $\widetilde{\Omega^{n-1}}(M, \omega)$. To achieve this we have to extend $\widetilde{\Omega^{n-1}}(M, \omega)$ to an n-term differential graded vector space. This is achieved by using the de Rham complex. We set $\tilde{L}_0 := \widetilde{\Omega^{n-1}}(M, \omega)$ and $\tilde{L}_i := \Omega^{(n-1+i)}(M)$ for $i \in \{-(n-1), ..., -1\}$. We turn $\tilde{L}(M, \omega) := \bigoplus_{i=-n+1}^{0} \tilde{L}_i$ into a differential graded vector space with differential $\tilde{l}_1$ by setting $\tilde{l}_1 = d : \Omega^k(M) \to \Omega^{k+1}(M)$ for $k \leq n-3$ and $\tilde{l}_1(\alpha) = (0, d\alpha) \in \widetilde{\Omega^{n-1}}(M, \omega)$ for $\alpha \in \Omega^{n-2}(M)$. Next we define a grounded Lie n-algebra on $\tilde{L}$ by setting $\tilde{l}_2((v, \alpha), (w, \beta)) = ([v, w], \iota_w\iota_v\omega)$ and

$$\tilde{l}_k((v_1, \alpha_1), .., (v_k, \alpha_k)) = -(-1)^{\frac{k(k+1)}{2}} \iota_{v_k}\iota_{v_{k-1}}...\iota_{v_1}\omega,$$

for $2 < k \leq n+1$. Before we prove that this construction yields a grounded Lie n-algebra we check that the map $\tilde{l}_2$ is well-defined. We have $d\iota_w\iota_v\omega = \mathcal{L}_w\iota_v\omega - \iota_w d\iota_v\omega = \mathcal{L}_w\iota_v\omega = -\iota_{[v,w]}\omega + \iota_v\mathcal{L}_w\omega = -\iota_{[v,w]}\omega$, which is equal to $-\omega^\flat([v, w])$ so that $([v, w], \iota_w\iota_v\omega)$ is indeed an element of $\widetilde{\Omega^{n-1}}(M, \omega)$.

Lemma 3.3. The graded vector space $\tilde{L}$ together with the multilinear maps $\{\tilde{l}_k \mid 1 \leq k \leq n+1\}$ forms a grounded Lie n-algebra.

Proof. It remains to be shown that the generalized Jacobi identity holds. Following Lemma 2.36 this can be reduced to checking the conditions (12) and (13). As $\tilde{l}_1(\alpha) = (0, d\alpha)$ and all higher brackets only depend on the first component, Equation (13) is trivial. Thus it remains to show that

$$-(-1)^{\frac{k(k+1)}{2}} d\iota_{v_k}\iota_{v_{k-1}}...\iota_{v_1}\omega + \sum_{1\leq i<j\leq k} (-1)^{i+j+1}(-1)(-1)^{\frac{(k-1)k}{2}} \omega([v_i, v_j], v_1, ..., \hat{v}_i, ..., \hat{v}_j, ..., v_k, \cdot, ..., \cdot) = 0.$$

This equation is equivalent to:

$$d\iota_{v_k}\iota_{v_{k-1}}...\iota_{v_1}\omega = (-1)^k \sum_{1\leq i<j\leq k} (-1)^{i+j}\omega([v_i, v_j], v_1, ..., \hat{v}_i, ..., \hat{v}_j, ..., v_k, \cdot, ..., \cdot) \tag{16}$$

In order to verify this equation we first apply Cartans formula $\mathcal{L}_v\mu = \iota_v d\mu + d\iota_v\mu$ to the left hand side, where $\mathcal{L}$ denotes the Lie derivative. We get:

$$d\iota_{v_k}\iota_{v_{k-1}}...\iota_{v_1}\omega = -\iota_{v_k}d(\iota_{v_{k-1}}...\iota_{v_1}\omega) + \mathcal{L}_{v_k}(\iota_{v_{k-1}}...\iota_{v_1}\omega).$$

Next we use the formula $\mathcal{L}_v\iota_w\mu = \iota_w\mathcal{L}_v\mu + \iota_{[v,w]}\mu$ to switch iteratively places between the contractions and the Lie derivative:

$$d\iota_{v_k}\iota_{v_{k-1}}...\iota_{v_1}\omega = -\iota_{v_k}d(\iota_{v_{k-1}}...\iota_{v_1}\omega) + \iota_{v_{k-1}}...\iota_{v_1}\mathcal{L}_{v_k}\omega - \sum_{i=1}^{k-1}\omega(v_1, ..., v_{i-1}, [v_i, v_k], v_{i+1}, ..., v_{k-1}, \cdot, ..., \cdot).$$

Next, again with Cartans formula, we observe that $\mathcal{L}_{v_k}\omega = \iota_{v_k}d\omega + d\iota_{v_k}\omega = 0 + d(-d\alpha_k) = 0$ and we get:

$$d\iota_{v_k}\iota_{v_{k-1}}...\iota_{v_1}\omega = -\iota_{v_k}d(\iota_{v_{k-1}}...\iota_{v_1}\omega) - \sum_{i=1}^{k-1}\omega(v_1, ..., v_{i-1}, [v_i, v_k], v_{i+1}, ..., v_{k-1}, ..., v_{k-1}, \cdot, ..., \cdot).$$

When we move $[v_i, v_k]$ to the front we have to move it past $i-1$ entries, so we get:

$$d\iota_{v_k}\iota_{v_{k-1}}...\iota_{v_1}\omega = -\iota_{v_k}d(\iota_{v_{k-1}}...\iota_{v_1}\omega) + \sum_{i=1}^{k-1}(-1)^i\omega([v_i, v_k], v_1, ..., \hat{v}_i, ..., v_{k-1}, \cdot, ..., \cdot) \tag{17}$$

Now we note, that $i \equiv i + k + k \mod 2$ so that we can rewrite the sum in the right hand side to obtain:

$$d\iota_{v_k}\iota_{v_{k-1}}...\iota_{v_1}\omega = -\iota_{v_k}d(\iota_{v_{k-1}}...\iota_{v_1}\omega) + (-1)^k\sum_{i=1}^{k-1}(-1)^{i+k}\omega([v_i, v_k], v_1, ..., \hat{v}_i, ..., v_{k-1}, \cdot, ..., \cdot).$$

Now we can show inductively that equation (16) holds. Assume that it holds for $k-1$, then we obtain, upon rewriting the first summand of the preceeding right hand side:

$$\begin{aligned} d\iota_{v_k}\iota_{v_{k-1}}...\iota_{v_1}\omega = -(-1)^{k-1} &\sum_{1\leq i<j\leq k-1}(-1)^{i+j}\omega([v_i, v_j], v_1, ..., \hat{v}_i, ..., \hat{v}_j, ..., v_k, \cdot, ..., \cdot) \\ &+(-1)^k\sum_{i=1}^{k-1}(-1)^{i+k}\omega([v_i, v_k], v_1, ..., \hat{v}_i, ..., v_{k-1}, \cdot, ..., \cdot) \\ = (-1)^k &\sum_{1\leq i<j\leq k}(-1)^{i+j}\omega([v_i, v_j], v_1, ..., \hat{v}_i, ..., \hat{v}_j, ..., v_k, \cdot, ..., \cdot). \end{aligned}$$

So we just have to show that the equation holds in the case $k = 1$. Then on the right hand side we sum over $1 \leq i < j \leq 1$, so that the right hand side is zero. On the left hand side we get $d\iota_{v_1}\omega = d(-d\alpha_1) = 0$. So Equation (16) is correct for $k = 1$ and thus for all k and hence $\tilde{L}(M, \omega)$ is a grounded Lie n-algebra. □

Definition 3.4. Let (M, ω) be a pre-n-plectic manifold. We call $\tilde{L}(M, \omega)$ the *Lie n-algebra of observables* of (M, ω).

Our next step is to generalize the notion of non-degeneracy. Unfortunately, because of dimension issues, we can not require $\omega_p^\flat : T_pM \to \Lambda^n T_p^*M$ to be an isomorphism in the general setting. The best we can hope for, is that $\omega^\flat$ is injective.

Definition 3.5. Let (M, ω) be a pre-n-plectic manifold. The couple (M, ω) is called *n-plectic* if $\omega^\flat$ is injective. We then call ω *non-degenerate.*

As in the symplectic case non-degeneracy of $\omega^\flat$ implies, that the map $\pi_\Omega : \widetilde{\Omega^{n-1}}(M, \omega) \to \Omega^{n-1}(M)$ is injective. Then $\widetilde{\Omega^{n-1}}(M, \omega)$ is isomorphic to $Im(\pi_\Omega) =: \Omega^{n-1}_{Ham}(M, \omega) \subset \Omega^{n-1}(M)$ and we can substitue $\widetilde{\Omega^{n-1}}(M, \omega)$ with $\Omega^{n-1}_{Ham}(M, \omega)$ in our definition of the Lie n-algebra. The differential from $\Omega^{n-2}(M)$ to $\Omega^{n-1}_{Ham}(M, \omega)$ is then the ordinary de Rham differential. We do not have to change anything in the definition of the other brackets, as the injectivity of $\omega^\flat$ guarantees, that for every $\alpha \in \Omega^{n-1}_{Ham}(M, \omega)$ there exists a unique vector field v_α such that $d\alpha = -\omega^\flat(v_\alpha)$. This Lie n-algebra is called $L(M, \omega)$ and the differential forms in $\Omega^{n-1}_{Ham}(M, \omega)$ are called *Hamiltonian forms.*

Definition 3.6. Let (M, ω) be n-plectic. We define the Lie n-algebra $L(M, \omega)$ as the graded vector space $L(M, \omega)_k := \tilde{L}(M, \omega)_k$ for $k \neq 0$ and $L(M, \omega)_0 := \Omega^{n-1}_{Ham}(M, \omega)$ with the bracket structure induced from $\widetilde{\Omega^{n-1}}(M, \omega)$. We call $L(M, \omega)$ the *algebra of observables of* (M, ω) in this case.

Remark 3.7. The map $\omega^\flat$ induces an isomorphism f of the differential graded vector spaces $L(M, \omega)$ and $\tilde{L}(M, \omega)$ by setting $f(\eta) = \eta$ for $\eta \in \bigoplus_{k=-n+1}^{-1} L(M, \omega)_k$ and $f(\alpha) = (v_\alpha, \alpha)$ for $\alpha \in \Omega^{n-1}_{Ham}(M, \omega)$. With this isomorphism the Lie n-algebra structure of $\tilde{L}(M, \omega)$ can be transported to $L(M, \omega)$. This transported Lie n-algebra structure turns f into a strict isomorphism of L_∞-algebras.

Let us discuss a few examples of n-plectic manifolds.

Example 3.8. A symplectic manifold (M, ω) is a 1-plectic manifold and $L(M, \omega) = (C^\infty(M), \{\cdot, \cdot\})$ is its Lie algebra of observables.

Example 3.9. Let $n \geq 2$ and (M, ν) be an oriented n-manifold with volume form ν. Then ν is closed and non-degenerate by definition and (M, ν) is an (n-1)-plectic manifold.

Example 3.10. Let Q be an m-manifold and $M = \Lambda^n T^*Q$ the n-th exterior power of its cotangent bundle with $1 \leq n \leq m$. We define $\theta \in \Omega^n(M)$ by

$$\theta_\eta(\xi_1, ..., \xi_n) = (T_\eta\pi)^*\eta(\xi_1, ..., \xi_n) = \eta(T_\eta\pi(\xi_1), ..., T_\eta\pi(\xi_n))$$

for $\eta \in M = \Lambda^n T^*Q$, $\xi_1, ..., \xi_n \in T_\eta M$ and $T\pi : T(\Lambda^n T^*Q) \to TQ$ the differential of the map $\pi : \Lambda^n T^*Q \to Q$. Setting $\omega = -d\theta$ we turn (M, ω) into an n-plectic manifold. In the case of $n = 1$ we retrieve the canonical symplectic structure on a cotangent bundle. To show non-degeneracy let us regard the situation in local coordinates:

- Let $x = (x_1, ..., x_m) : U \subset Q \to \mathbb{R}^m$ be a coordinate system with $x_i \in C^\infty(U, \mathbb{R})$

- $dx_1, ..., dx_m \in \Omega^1(U) = \Gamma(U, T^*U)$ form a basis of T_a^*U at any point $a \in U$. Thus we get a coordinate system for T^*U, where we define $p_i : T^*U \to \mathbb{R}$ to be $p_i(\alpha) = \alpha_i$, by mapping $\alpha = \sum_{i=1}^m \alpha_i dx_i|_a \in T_a^*Q$ to $(x_1(a), ..., x_m(a), p_1(\alpha), ..., p_m(\alpha))$.

- Accordingly for $1 \le i_1 < ... < i_n \le m$ the $(dx_{i_1} \wedge ... \wedge dx_{i_n})|_a =: dx^I|_a$ form a basis of $\Lambda^n T_a^*U$ and we get a coordinate system for $\Lambda^n T^*U$ by

$$\alpha = \sum_I \alpha_I dx^I|_a \quad \mapsto \quad ((x_i(a))_i, (p_I(\alpha))_I) = ((x_i(\pi(\alpha)))_i, (p_I(\alpha))_I),$$

where i runs from 1 to m, I runs through all strictly increasing multi-indices of length n and $p_I(\alpha) = \alpha_I$. Every single p_I is a map from $\Lambda^n T^*U$ to $\mathbb{R}$.

- Now we consider $d\theta_\bullet = d(T_\bullet\pi)^*(\sum p_I(\bullet)dx^I) = \sum(d(p_I(\bullet)) \wedge (T_\bullet\pi)^*dx^I)$. As $(\tilde{x}_i, P_I)$ with $\tilde{x}_i = x_i \circ \pi$ forms a chart for $\Lambda^n T^*U$, the $\{d\tilde{x}_i\}$ together with the $\{dP_I\}$ form a basis of $T_\eta^*(\Lambda^n T^*U)$ for $\eta \in \Lambda^n T^*U$. We can also regard its dual basis consisting of $\frac{\partial}{\partial \tilde{x}_i}$ and $\frac{\partial}{\partial p_I}$. To show that $d\theta$ is non-degenerate we have to show, that for any $0 \neq \xi = \xi_0 \in T_\eta(\Lambda^n T^*U)$, $d\theta_\eta(\xi_0, \cdot, ..., \cdot)$ is not zero, i.e., there exist $\xi_1, ..., \xi_n$ such that $d\theta(\xi_0, \xi_1, ..., \xi_n) \neq 0$. First let us assume that $dp_I(\xi_0) \neq 0$ for some $I = (i_1, ..., i_n)$. Then we just pick $\xi_k = \frac{\partial}{\partial \tilde{x}_k}$ to guarantee that $d\theta(\xi_0, ..., \xi_n)$ is not zero. If $dp_I(\xi_0) = 0$ for all I, we can without loss of generality assume that $d\tilde{x}_1(\xi_0)$ is non-zero. Then we pick some multi-index $I = (i_1, ...i_n)$ with $i_1 = 1$ and set $\xi_k = \frac{\partial}{\partial \tilde{x}_{i_k}}$ for $k > 1$ and $\xi_1 := \frac{\partial}{\partial p_I}$ to guarantee that $d\theta(\xi_0, ..., \xi_n)$ is non-zero. Thus $d\theta$ (and equivalently $-d\theta$) is non-degenerate and thus ω is an n-plectic form.

Example 3.11. Let G be a connected semi-simple Lie group. We will construct a 2-plectic form on G using the following facts:

- The Lie bracket is Ad_g-equivariant for all $g \in G$. As G is semi-simple, we have $[\mathfrak{g}, \mathfrak{g}] = \mathfrak{g}$.

- The (symmetric) Killing-form $\langle \cdot, \cdot \rangle : \mathfrak{g} \times \mathfrak{g} \to \mathbb{R}$ is Ad_g-invariant for all $g \in G$ and ad_X is a skew-adjoint linear map for all $X \in \mathfrak{g}$. It is non-degenerate, and negative definite for compact semi-simple Lie groups.

- The Maurer-Cartan 1-form $\theta^L \in \Omega^1(G, \mathfrak{g})$ defined by $\theta^L_g = T_g(L_{g^{-1}}) : T_gG \to T_eG = \mathfrak{g}$, where $L_g : G \to G$ is the left multiplication by g, is by construction left-invariant.

We define $\omega \in \Omega^3(G)$ by $\omega(u, v, w) = \langle \theta^L(u), [\theta^L(v), \theta^L(w)] \rangle$. Non-degeneracy follows from $[\mathfrak{g}, \mathfrak{g}] = \mathfrak{g}$ and the non-degeneracy of the Killing form. The left-invariance of θ^L implies that ω, too, is left-invariant. Using the description $Ad_g = T(L_g) \circ T(R_{g^{-1}})$, the Ad_g-invariance of the Killing form and the Ad-equivariance of the Lie bracket one can also show, that ω is right-invariant. Any bi-invariant form on a Lie group is automatically closed, so ω is in $\Omega^3_{cl}(G)$ and non-degenerate and thus defines a 2-plectic structure on G.

4 Symmetries of pre-n-plectic manifolds

4.1 Weakly Hamiltonian group actions

Having constructed a notion of observables for n-plectic manifolds we will now try to generalize the notions of Hamiltonian group action and co-moment maps. Let us first recall the classical situation (from the perspective of co-moment maps):

Definition 4.1. Let (M, ω) be a (pre-)symplectic manifold and $\mathfrak{X}(M, \omega)$ the sub Lie algebra of vector fields v satisfying $\mathcal{L}_v\omega = 0$. Let $\pi_{\mathfrak{X}} : \widetilde{C^\infty}(M, \omega) \to \mathfrak{X}(M, \omega) \subset \mathfrak{X}(M)$, $(v, f) \mapsto v$ denote the canonical projection. Let $\vartheta : M \times G \to M$ be a right action of a connected group G and $\zeta : \mathfrak{g} \to \mathfrak{X}(M)$ the associated infinitesimal action.

1. The action ϑ is called *(pre-)symplectic* if $\vartheta_g^*\omega = \omega$ for all $g \in G$ or equivalently when $Im(\zeta) \subset \mathfrak{X}(M, \omega)$.
2. A (pre-)symplectic ϑ is called *(weakly) Hamiltonian* if there exists a linear lift $j : \mathfrak{g} \to \widetilde{C^\infty}(M, \omega)$ along $\pi_{\mathfrak{X}}$ (i.e. fulfilling $\pi_{\mathfrak{X}} \circ j = \zeta$).
3. A Hamiltonian group action ϑ is called *strongly Hamiltonian* if there exists a lift $j : \mathfrak{g} \to \widetilde{C^\infty}(M, \omega)$ which is also a Lie algebra homomorphism. The map j is then called a *strong co-moment map* for the strongly Hamiltonian group action.

Remark 4.2. The group G is required to be connected to guarantee the equivalence of $\vartheta_g^*\omega = \omega$ for all $g \in G$ and $Im(\zeta) \subset \mathfrak{X}(M, \omega)$. We have chosen right actions in order to assure that the infinitesimal action ζ is a Lie algebra homomorphism, for left actions it would be an anti-homomorphism. Furthermore $\pi_{\mathfrak{X}}$ is always a Lie algebra homomorphism in this situation.

Now we will generalize the concept of co-moment maps to the (pre-)n-plectic case and call the generalized construction homotopy co-moment map. To do that, let us first observe that, by definition, $\pi_{\mathfrak{X}}\pi_0 : \tilde{L}(M, \omega) \overset{\pi_0}{\to} \widetilde{\Omega^{n-1}}(M, \omega) \overset{\pi_{\mathfrak{X}}}{\to} \mathfrak{X}(M)$ is a Lie n-algebra homomorphism. We can, as in the (pre-)symplectic case, define the sub Lie algebra of vector fields satisfying $\mathcal{L}_v\omega = 0$ and denote it by $\mathfrak{X}(M, \omega)$. It is a Lie sub-algebra because $\mathcal{L}_{[v,w]} = \mathcal{L}_v\mathcal{L}_w - \mathcal{L}_w\mathcal{L}_v$. It holds that $Im(\pi_{\mathfrak{X}}\pi_0) \subset \mathfrak{X}(M, \omega)$, because for $(v, \alpha) \in \widetilde{\Omega^{n-1}}(M, \omega)$ we have $L_v\omega = \iota_v d\omega + d\iota_v\omega = 0 - dd\alpha = 0$. So let us first define the generalization of symplectic actions:

Definition 4.3. Let (M, ω) be (pre-)n-plectic. A right action $\vartheta : M \times G \to M$ of some Lie group G is called *(pre-)n-plectic* if $\vartheta_g^*(\omega) = \omega$ for all $g \in G$.

Lemma 4.4. Let G be connected. Then ϑ is (pre-)n-plectic if and only if $Im(\zeta) \subset \mathfrak{X}(M, \omega)$.

Proof. We have to show, that ω is invariant exactly if $\mathcal{L}_{\zeta(X)}\omega = 0$ for all $X \in \mathfrak{g}$. First we differentiate the expression $(\vartheta_{exp(tX)})^*\omega$ and use that the flow $\vartheta_{exp(tX)}$ generates the vector field $\zeta(X)$ to obtain

$$\left.\frac{d}{dt}\right|_{t=0} (\vartheta_{exp(tX)})^*\omega = \mathcal{L}_{\zeta(X)}\omega.$$

Thus if ω is G-invariant $(\vartheta_{exp(tX)})^*\omega$ is constant and thus $L_{\zeta(X)}\omega = \frac{d}{dt}\big|_{t=0}(\vartheta_{exp(tX)})^*\omega = 0$ for all $X \in \mathfrak{g}$. On the other hand if $\mathcal{L}_{\zeta(X)}\omega = 0$ for all X then $\frac{d}{dt}\big|_{t=0}(\vartheta_{exp(tX)})^*\omega = 0$ and $(\vartheta_g)^*\omega$ is constant for $g \in exp(U) \subset G$ for some open neighbourhood U of zero in $\mathfrak{g}$. The set $I = \{g \mid \vartheta_g^*\omega = \omega\}$ is closed by definition, but for every $g \in I$ in this set $g \cdot exp(\mathfrak{g})$, which contains an open neighbourhood of g is in I, too. So I is open and closed, and thus for a connected G we have $I = G$. Thus ω is invariant under G. □

Remark 4.5. Note that $d\iota_{\zeta(X)}\omega = -\iota_{\zeta(X)}d\omega + \mathcal{L}_{\zeta(X)}\omega = \mathcal{L}_{\zeta(X)}\omega$, i.e. $\iota_{\zeta(X)}\omega$ is closed for all X if and only if the action is pre-n-plectic.

Next let us define generalized Hamiltonian actions:

Definition 4.6. A (pre-)n-plectic action ϑ is called *(weakly) Hamiltonian* if there exists a linear lift j of ζ along $\pi_{\mathfrak{X}} \circ \pi_0$, where $\pi_0 : \tilde{L}(M,\omega) \to \tilde{L}(M,\omega)_0$ is the projection. Then we call j a *(weak) co-moment map (for the action ϑ)*.

Note that π_0 is a projection on a summand and thus surjective. So we can equivalently require the existence of a lift along $\pi_{\mathfrak{X}}$. This means that an action is Hamiltonian iff the image of ζ lies in $\pi_{\mathfrak{X}}(\widetilde{\Omega^{n-1}}(M,\omega)) = \{v \in \mathfrak{X}(M,\omega) \mid \exists\, \alpha : \iota_v\omega = -d\alpha\}$ i.e. $\iota_{\zeta(X)}\omega$ has to be exact for all X. Thus (in analogy with the classical case) we can regard the situation in terms of the following diagram:

$$\begin{array}{ccccc} \widetilde{\Omega^{n-1}}(M,\omega) & \xrightarrow{\pi_{\mathfrak{X}}} & \mathfrak{X}(M,\omega) & \xrightarrow[X\mapsto[\iota_X\omega]]{\gamma} & H^n_{dR}(M) \\ & \overset{j}{\nwarrow} & \uparrow \zeta & & \\ & & \mathfrak{g} & & \end{array}$$

Remark 4.5 guarantees that for a (pre-)n-plectic group action the map γ is well-defined and the top row is exact. Then ζ can be lifted to j if and only if $\gamma\zeta = 0$. The kernel of $\pi_{\mathfrak{X}}$ consists of all pairs (v,α) where v is zero i.e. elements of the form $(0,\alpha)$, with $d\alpha = 0$. So we can extend the diagram to the following one, where the top row stays exact.

$$\begin{array}{ccccccccc} 0 & \longrightarrow & \Omega^{n-1}_{cl}(M) & \xrightarrow{(0,id)} & \widetilde{\Omega^{n-1}}(M,\omega) & \xrightarrow{\pi_{\mathfrak{X}}} & \mathfrak{X}(M,\omega) & \xrightarrow[X\mapsto[\iota_X\omega]]{\gamma} & H^n_{dR}(M) \\ & & & & & \overset{j}{\nwarrow} & \uparrow \zeta & & \\ & & & & & & \mathfrak{g} & & \end{array} \qquad (18)$$

Given a linear j we can ask ourselves, whether it can be extended/altered to an L_∞-homomorphism. To do that, let us first extend the above diagram to a diagram of L_∞-algebras. For that we substitute $\Omega^{n-1}_{cl}(M)$ with the chain complex $C^\infty(M) \xrightarrow{d} \Omega^1(M)... \to .. \to \Omega^{n-2}(M) \xrightarrow{d} \Omega^{n-1}_{cl}(M)$ and denote the latter by $\Omega^{\leq n-1}_{cl}(M)$. We give it the structure of a grounded Lie n-algebra by setting all brackets beyond $\tilde{l}_1$ to be zero. We extend the map from $\Omega^{n-1}_{cl}(M)$ to $\widetilde{\Omega^{n-1}}(M,\omega)$ by identities to get an injective map $i : \Omega^{\leq n-1}_{cl}(M) \to \tilde{L}(M,\omega)$. Then we turn $H^n_{dR}(M)$ into a Lie algebra with the trivial bracket and substitute $\pi_{\mathfrak{X}}$ with $q = \pi_{\mathfrak{X}}\pi_0$, where π_0 is the projection onto $\tilde{L}(M,\omega)_0$. In this setting the above considerations take the following form:

Lemma 4.7. Let (M, ω) be a pre-n-plectic manifold. The following sequence of L_∞-algebras is exact:

$$0 \longrightarrow \Omega^{\leq n-1}_{cl}(M) \xrightarrow{\ i\ } \tilde{L}(M, \omega) \xrightarrow{\ q\ } \mathfrak{X}(M, \omega) \xrightarrow{\ \gamma\ } H^n_{dR}(M)$$

A Lie algebra homomorphism $\zeta : \mathfrak{g} \to \mathfrak{X}(M, \omega)$ can be lifted to a linear map $j : \mathfrak{g} \to \tilde{L}(M, \omega)$ if and only if $\gamma\zeta = 0$.

Proof. The only thing that remains to be checked is that γ is a Lie algebra morphism. This follows from the following calculation:

$$\iota_{[v,w]}\omega = \mathcal{L}_v \iota_w \omega - \iota_w \mathcal{L}_v \omega = d\iota_v \iota_w \omega + \iota_v d\iota_w \omega = d\iota_v \iota_w \omega.$$

The last equation follows from the fact that $d\iota_w \omega = \mathcal{L}_w \omega - \iota_w d\omega = 0$. So we have shown that the de Rham class $[\iota_{[v,w]}\omega]_{dR} = 0$ and thus the lemma holds. □

Remark 4.8. If $H^n_{dR}(M) = 0$ or $[\mathfrak{g}, \mathfrak{g}] = \mathfrak{g}$ then any pre-n-plectic action is Hamiltonian. The first case is clear, for the second case one just exploits the fact that γ is a Lie algebra homomorphism onto the zero bracket. If $[\mathfrak{g}, \mathfrak{g}] = \mathfrak{g}$ then any X can be realized by $\sum \lambda_i [A_i, B_i]$ and

$$\gamma\zeta(X) = \gamma\zeta(\sum \lambda_i [A_i, B_i]) = \sum \lambda_i [\gamma\zeta(A_i), \gamma\zeta(B_i)] = \sum \lambda_i \cdot 0 = 0.$$

Definition 4.9. Let (M, ω) be pre-n-plectic, G a connected Lie group and $\vartheta : M \times G \to M$ a pre-n-plectic right action with associated $\zeta : \mathfrak{g} \to \mathfrak{X}(M, \omega)$.

1. We call $F = \{f_k\} : \mathfrak{g} \to \tilde{L}(M, \omega)$ a *strong homotopy co-moment map (for the action ϑ)* if $\pi_{\mathfrak{X}} \circ F = \zeta$ and F is an L_∞-morphism. We also call -by a slight abuse of language- the component $j = f_1 : \mathfrak{g} \to \tilde{L}_0 = \widetilde{\Omega^{n-1}}(M, \omega)$ of F a *homotopy co-moment map*.

2. We call ϑ *strongly Hamiltonian* if there exists a strong homotopy co-moment map (for the ζ defined by ϑ).

Remark 4.10. In both cases (i.e. for F and for j as above) we sometimes only refer to a "co-moment map" omitting the word "homotopy".

4.2 Strongly Hamiltonian group actions

Let us now turn to the question whether/when a Hamiltonian group action is strongly Hamiltonian i.e. when j can be changed/extended into an L_∞-morphism. For that let us reformulate the conditions on our morphisms into the language of the Chevalley-Eilenberg complex. In this section $\mathfrak{g}$ will always denote a finite-dimensional real Lie algebra with Lie bracket $[\cdot, \cdot]$.

- For a vector space V, let $\delta : \Lambda^k(\mathfrak{g}, V) := Hom(\Lambda^k \mathfrak{g}, V) = \Lambda^k \mathfrak{g}^* \otimes V \to \Lambda^{k+1}(\mathfrak{g}, V)$ denote the Chevalley-Eilenberg differential with respect to the trivial representation of $\mathfrak{g}$ on V.

- Let $\partial : \Lambda^k(\widetilde{\Omega^{n-1}}(M,\omega), V) = Hom(\Lambda^k(\widetilde{\Omega^{n-1}}(M,\omega)), V) \to \Lambda^{k+1}(\widetilde{\Omega^{n-1}}(M,\omega), V)$ denote the natural analogue of the Chevalley-Eilenberg differential with respect to the trivial representation and the bracket $\tilde{l}_2$ given by the formula:

$$(\partial\phi)(x_1, ..., x_{k+1}) = \sum_{1\leq i<j\leq n+1} (-1)^{i+j}\phi(\tilde{l}_2(x_i, x_j), x_1, ..., \hat{x}_i, ..., \hat{x}_j, ..., x_{k+1}),$$

  for $\phi : \Lambda^k(\widetilde{\Omega^{n-1}}(M,\omega)) \to V$ and $x_1, ..., x_{k+1} \in \widetilde{\Omega^{n-1}}(M,\omega)$.

Using this notation we can easily check the following equivalences:

- The conditions for the case of grounded Lie n-algebras from Lemma 2.35 can be rewritten as $\tilde{l}_1\tilde{l}_k = \partial\tilde{l}_{k-1}$ for $k \in \{1, ..., n+2\}$ and $\tilde{l}_k(\tilde{l}_1(x_1), x_2..., x_k) = 0$ (where $\tilde{l}_0$ and $\tilde{l}_{n+2}$ are set to be zero).

- A family of maps $f_1, ..., f_n$ forms an L_∞-morphism (as described in Lemma 2.36) from $\mathfrak{g}$ to $\tilde{L}(M,\omega)$ if and only if $-\delta f_{k-1} = \tilde{l}_1 f_k + f_1^*\tilde{l}_k$ for $k \in \{2, ..., n+1\}$ (where f_{n+1} is set to be zero).

Furthermore, if $j : \mathfrak{g} \to \tilde{L}(M,\omega)_0$ is a weak co-moment, then we have $j^*\partial\tilde{l}_k = \delta j^*\tilde{l}_k$ for $1 \leq k \leq n+1$. To see that, we first observe that $\pi_{\mathfrak{X}}(\tilde{l}_2(j(X), j(Y)) - j([X,Y])) = 0$ for all $X, Y \in \mathfrak{g}$ as j is a linear lift of ζ. Note that this furthermore implies that $\pi_\omega(\tilde{l}_2(j(X), j(Y)) - j([X,Y]))$ is closed. Then we calculate

$$(j^*\tilde{\partial}l_k - \delta j^*\tilde{l}_k)(X_1, ..., X_{k+1}) = \sum_{i<j}(-1)^{i+j}\tilde{l}_k(\tilde{l}_2(j(X_i), j(X_j)) - j([X_i, X_j]), X_1, ..., \hat{X}_i, ..., \hat{X}_j, ..., X_{k+1}).$$

The assertion $j^*\partial\tilde{l}_k = \delta j^*\tilde{l}_k$ then follows from the fact that $\tilde{l}_k(x_1, ..., x_k)$ only depends on the vector field component of the elements $x_1, ..., x_k$.

Using this we can define the obstruction classes preventing a Hamiltonian action from being strongly Hamiltonian.

Lemma 4.11. Let j be a weak co-moment map for an infinitesimal Hamiltonian group action $\zeta : \mathfrak{g} \to \mathfrak{X}(M,\omega)$ on a pre-n-plectic manifold. Then for $k \in \{1, ..., n\}$ the following classes are well-defined

$$h_k = [[j^*\tilde{l}_{k+1}]_{CE}]_{dR} \in H^{k+1}(\mathfrak{g}, H^{n-k}_{dR}(M)) = H^{k+1}(\mathfrak{g}) \otimes H^{n-k}_{dR}(M),$$

where $[\cdot]_{dR}$ means taking the de Rham equivalence class and $[\cdot]_{CE}$ means taking the Chevalley-Eilenberg equivalence class.

Proof. First assume $k \neq 1$. Let us regard the below exact sequence (where $(\Lambda^k\mathfrak{g}^*)_{cl}$ resp. $(\Lambda^k\mathfrak{g}^*)_{ex}$ denotes δ-closed resp. δ-exact elements of $\Lambda^k\mathfrak{g}^*$) and its exact subsequence:

$$\begin{array}{ccccccc}
0 & \longrightarrow & (\Lambda^{k+1}\mathfrak{g}^*)_{cl} & \xrightarrow{i} & \Lambda^{k+1}\mathfrak{g}^* & \xrightarrow{\delta} & \Lambda^{k+2}\mathfrak{g}^* \\
 & & \uparrow{\scriptstyle i} & & \uparrow{\scriptstyle i} & & \uparrow{\scriptstyle i} \\
0 & \longrightarrow & (\Lambda^{k+1}\mathfrak{g}^*)_{ex} & \xrightarrow{i} & (\Lambda^{k+1}\mathfrak{g}^*)_{ex} & \xrightarrow{\delta} & 0
\end{array}$$

By dividing out the subsequence we get the following exact(!) sequence

$$0 \longrightarrow \frac{(\Lambda^{k+1}\mathfrak{g}^*)_{cl}}{(\Lambda^{k+1}\mathfrak{g}^*)_{ex}} \xrightarrow{i} \frac{\Lambda^{k+1}\mathfrak{g}^*}{(\Lambda^{k+1}\mathfrak{g}^*)_{ex}} \xrightarrow{\delta} \Lambda^{k+2}\mathfrak{g}^*$$

Analogously we get the following exact sequence by dividing $0 \to \Omega^{n-k}_{ex}(M) \xrightarrow{i} \Omega^{n-k}_{ex}(M) \to 0$ out of $0 \to \Omega^{n-k}_{cl}(M) \xrightarrow{i} \Omega^{n-k}(M) \xrightarrow{\tilde{l}_1} \Omega^{n-k+1}(M)$:

$$0 \longrightarrow \frac{\Omega^{n-k}_{cl}(M)}{\Omega^{n-k}_{ex}(M)} \xrightarrow{i} \frac{\Omega^{n-k}(M)}{\Omega^{n-k}_{ex}(M)} \xrightarrow{\tilde{l}_1} \Omega^{n-k+1}(M)$$

Since we are working over a field taking tensor products preserves exactness and we arrive at the following commutative diagram with exact rows and columns:

$$\begin{array}{ccccccc}
 & & 0 & & 0 & & 0 \\
 & & \downarrow & & \downarrow & & \downarrow \\
0 & \longrightarrow & \frac{(\Lambda^{k+1}\mathfrak{g}^*)_{cl}}{(\Lambda^{k+1}\mathfrak{g}^*)_{ex}} \otimes \frac{\Omega^{n-k}_{cl}(M)}{\Omega^{n-k}_{ex}(M)} & \longrightarrow & \frac{\Lambda^{k+1}\mathfrak{g}^*}{(\Lambda^{k+1}\mathfrak{g}^*)_{ex}} \otimes \frac{\Omega^{n-k}_{cl}(M)}{\Omega^{n-k}_{ex}(M)} & \longrightarrow & \Lambda^{k+2}\mathfrak{g}^* \otimes \frac{\Omega^{n-k}_{cl}(M)}{\Omega^{n-k}_{ex}(M)} \\
 & & \downarrow & & \downarrow & & \downarrow \\
0 & \longrightarrow & \frac{(\Lambda^{k+1}\mathfrak{g}^*)_{cl}}{(\Lambda^{k+1}\mathfrak{g}^*)_{ex}} \otimes \frac{\Omega^{n-k}(M)}{\Omega^{n-k}_{ex}(M)} & \longrightarrow & \frac{\Lambda^{k+1}\mathfrak{g}^*}{(\Lambda^{k+1}\mathfrak{g}^*)_{ex}} \otimes \frac{\Omega^{n-k}(M)}{\Omega^{n-k}_{ex}(M)} & \longrightarrow & \Lambda^{k+2}\mathfrak{g}^* \otimes \frac{\Omega^{n-k}(M)}{\Omega^{n-k}_{ex}(M)} \\
 & & \downarrow & & \downarrow & & \downarrow \\
0 & \longrightarrow & \frac{(\Lambda^{k+1}\mathfrak{g}^*)_{cl}}{(\Lambda^{k+1}\mathfrak{g}^*)_{ex}} \otimes \Omega^{n-k+1}(M) & \longrightarrow & \frac{\Lambda^{k+1}\mathfrak{g}^*}{(\Lambda^{k+1}\mathfrak{g}^*)_{ex}} \otimes \Omega^{n-k+1}(M) & \longrightarrow & \Lambda^{k+2}\mathfrak{g}^* \otimes \Omega^{n-k+1}(M)
\end{array}$$

Now we interpret $j^*\tilde{l}_{k+1}$ as an element of $\Lambda^{k+1}\mathfrak{g}^* \otimes \Omega^{n-k}(M)$ and map it to $\frac{\Lambda^{k+1}\mathfrak{g}^*}{(\Lambda^{k+1}\mathfrak{g}^*)_{ex}} \otimes \frac{\Omega^{n-k}(M)}{\Omega^{n-k}_{ex}(M)}$ via the tensor product of the quotient maps. We will denote it by $[[j^*\tilde{l}_{k+1}]_{CE}]_{dR}$ to show that the exact part was divided out, though of course we have not yet verified that this element is closed. Indeed,

$$(\delta \otimes id)([[j^*\tilde{l}_{k+1}]_{CE}]_{dR} = [\delta j^*\tilde{l}_{k+1}]_{dR} = [j^*\partial\tilde{l}_{k+1}]_{dR} = [j^*\tilde{l}_1\tilde{l}_{k+2}]_{dR} = [\tilde{l}_1 j^*\tilde{l}_{k+2}]_{dR} = 0.$$

By the exactness of the columns, $[[j^*\tilde{l}_{k+1}]_{CE}]_{dR}$ is an element of $\frac{(\Lambda^{k+1}\mathfrak{g}^*)_{cl}}{(\Lambda^{k+1}\mathfrak{g}^*)_{ex}} \otimes \frac{\Omega^{n-k}(M)}{\Omega^{n-k}_{ex}(M)}$. Next we can calculate

$$(id \otimes \tilde{l}_1)([[j^*\tilde{l}_{k+1}]_{CE}]_{dR}) = [j^*\tilde{l}_1\tilde{l}_{k+1}]_{CE} = [j^*\partial\tilde{l}_k]_{CE} = [\delta j^*\tilde{l}_k]_{CE} = 0.$$

By the exactness of the rows, $h_k = [[j^*\tilde{l}_{k+1}]_{CE}]_{dR}$ is a unique well-defined element of the space $\frac{(\Lambda^{k+1}\mathfrak{g}^*)_{cl}}{(\Lambda^{k+1}\mathfrak{g}^*)_{ex}} \otimes \frac{\Omega^{n-k}_{cl}(M)}{\Omega^{n-k}_{ex}(M)} = H^{k+1}(\mathfrak{g}) \otimes H^{n-k}_{dR}(M)$.

For the case $k = 1$ we substitute the exact sequence $0 \to \Omega^{n-k}_{cl}(M) \to \Omega^{n-k}(M) \xrightarrow{\tilde{l}_1} \Omega^{n-k+1}(M)$ by the exact sequence $0 \to \Omega^{n-1}_{cl}(M) \to \widetilde{\Omega^{n-1}}(M,\omega) \xrightarrow{\pi_{\mathfrak{X}}} \mathfrak{X}(M,\omega)$. Then δ-closedness is shown analogously to the previous case $k \neq 1$. For $\tilde{l}_1$-closedness note that $[j^*\tilde{l}_2]_{CE} = [j^*\tilde{l}_2 + \delta j]_{CE} \in H^2(\mathfrak{g}) \otimes \Omega^{n-1}_{cl}(M)$ due to $\pi_{\mathfrak{X}}(j^*\tilde{l}_2 + \delta j) = 0$. Hence $[[j^*\tilde{l}_2]_{CE}]_{dR}$ is an element of $H^2(\mathfrak{g}) \otimes H^{n-1}_{dR}(M)$ which finishes the proof. □

Remark 4.12. The classes h_k do not depend on the choice of j as it only can vary by a map with values in $\Omega^{n-1}_{cl}(M)$ which are "ignored" by $\tilde{l}_k$ (cf. the discussion before Lemma 4.11).

Theorem 4.13. A Hamiltonian action is strongly Hamiltonian if and only if $h_k = 0$ for all $k \in \{1, ..., n\}$.

Proof. Let us first assume that the action is strongly Hamiltonian i.e. we have $f_1, ..., f_n$ satisfying $-\delta f_{k-1} = \tilde{l}_1 f_k + f_1^* \tilde{l}_k$. Then for $k \in \{2, ..., n+1\}$, $[\tilde{l}_1 f_k + f_1^* \tilde{l}_k]_{CE} = 0 \in H^k(\mathfrak{g}) \otimes \Omega^{n+1-k}(M)$. Now we divide out the de Rham equivalence relation and get $0 = [[\tilde{l}_1 f_k + f_1^* \tilde{l}_k]_{CE}]_{dR} = [[f_1^* \tilde{l}_k]_{CE}]_{dR} = [[j^* \tilde{l}_k]_{CE}]_{dR} = h_{k-1}$ so all the h_k are zero.

Assume now that $h_1, ..., h_n$ are zero and choose any weak co-moment j. We will construct maps $f_1, ..., f_n$ iteratively starting with f_n. In the sequel f_r and $\hat{f}_r$ are always elements of $\Lambda^r(\mathfrak{g}^*) \otimes \tilde{L}_{-r+1}$ for $r \in \{1, ..., n\}$. Of course $f_{n+1} := 0$.

Initial Step: $(k = n)$ Define $\hat{f}_n \in \Lambda^n(\mathfrak{g}^*) \otimes C^\infty_{cl}(M) \in$ as the δ-potential of $-j^*\tilde{l}_{n+1} \in \Lambda^{n+1}(\mathfrak{g}^*) \otimes H^0_{dR}(M) = \Lambda^{n+1}(\mathfrak{g}^*) \otimes C^\infty_{cl}(M)$. Its existence is assured by $[[-j^*\tilde{l}_{n+1}]_{dR}]_{CE} = [-j^*\tilde{l}_{n+1}]_{CE} = -h_n = 0$.

Iteration Step: $(k \in \{n-1, ..., 2\})$ Assume we already dispose of $f_n, ..., f_{k+2}$ and $\hat{f}_{k+1}$ such that $\delta \hat{f}_{k+1} = -(j^*\tilde{l}_{k+2} + \tilde{l}_1 f_{k+2})$.

Considering the element $-(j^*\tilde{l}_{k+1} + \tilde{l}_1 \hat{f}_{k+1}) \in \Lambda^{k+1}(\mathfrak{g}^*) \otimes \Omega^{n-k-1}(M)$, we obtain

$$\delta(-(j^*\tilde{l}_{k+1} + \tilde{l}_1 \hat{f}_{k+1})) = -(j^*\partial\tilde{l}_{k+1} + \tilde{l}_1(-(j^*\tilde{l}_{k+2} + \tilde{l}_1 f_{k+2}))) = -(j^*\tilde{l}_1\tilde{l}_{k+2} - j^*\tilde{l}_1\tilde{l}_{k+2}) = 0.$$

Thus we can interpret $[-(j^*\tilde{l}_{k+1} + \tilde{l}_1 \hat{f}_{k+1})]_{CE}$ as an element of $H^{k+1}(g) \otimes \Omega^{n-k}(M)$. Furthermore, we have

$$\tilde{l}_1([-(j^*\tilde{l}_{k+1} + \tilde{l}_1 \hat{f}_{k+1})]_{CE}) = [-j^*\tilde{l}_1\tilde{l}_{k+1}]_{CE} = -[j^*\partial l_k]_{CE} = -[\delta j^*\tilde{l}_k]_{CE} = 0.$$

Thus $[-(j^*\tilde{l}_{k+1} + \tilde{l}_1 \hat{f}_{k+1})]_{CE}$ is an element of $H^{k+1}(\mathfrak{g}) \otimes \Omega^{n-k}_{cl}(M)$. Using the exactness of the sequence

$$H^{k+1}(\mathfrak{g}) \otimes \Omega^{n-k-1}(M) \xrightarrow{id\otimes\tilde{l}_1} H^{k+1}(\mathfrak{g}) \otimes \Omega^{n-k}_{cl}(M) \xrightarrow{id\otimes[\cdot]_{dR}} H^{k+1}(\mathfrak{g}) \otimes H^{n-k}_{dr}(M)$$

and the fact that $[[-(j^*\tilde{l}_{k+1} + \tilde{l}_1\hat{f}_{k+1})]_{CE}]_{dR} = [[-j^*\tilde{l}_{k+1}]_{CE}]_{dR} = -h_k = 0$, we get a $\tilde{l}_1$-potential $[p_{k+1}]_{CE} \in H^{k+1}(\mathfrak{g}) \otimes \Omega^{n-k-1}(M)$ satisfying $\tilde{l}_1[p_{k+1}]_{CE} = [-(j^*\tilde{l}_{k+1} + \tilde{l}_1\hat{f}_{k+1})]_{CE}$. Let p_{k+1} be a Chevalley-Eilenberg-representative of this potential. We put $f_{k+1} = \hat{f}_{k+1} + p_{k+1}$ and obtain

$$[-(j^*\tilde{l}_{k+1} + \tilde{l}_1 f_{k+1})]_{CE} = 0 \in H^{k+1}(\mathfrak{g}) \otimes \Omega^{n-k}_{cl}(M).$$

Thus there exists a Chevalley-Eilenberg-potential of this element. Let $\hat{f}_k$ be such a potential.

Final step: (k=1) We assume we already have determined $f_n, ..., f_3$ and $\hat{f}_2$. Consider the element $[-(j^*\tilde{l}_2 + \tilde{l}_1\hat{f}_2 + \delta j)]_{CE}$. Applying $[\cdot]_{dR}$ turns it into zero, implying that it has an $\tilde{l}_1$-potential. Let p_2 be a Chevalley-Eilenberg-representative of it. We set $f_2 = \hat{f}_2 + p_2$. Now $[-(j^*\tilde{l}_2 + \tilde{l}_1 f_2 + \delta j)]_{CE} = 0 \in H^2(\mathfrak{g}) \otimes \Omega^{n-1}_{cl}(M)$. So there exists a δ-potential $c \in \mathfrak{g}^* \otimes \Omega^{n-1}_{cl}(M)$ of it. We set $f_1 := j + c$.

Verification: To finish the proof, we have to check that the constructed $f_1, ..., f_n$ form an L_∞-morphism. For $k \neq 1$ we get, upon observing that $c^*\tilde{l}_k = 0$:

$$-\delta f_k = -\delta(\hat{f}_k + p_k) = -\delta\hat{f}_k = j^*\tilde{l}_{k+1} + \tilde{l}_1 f_{k+1} = f_1^*\tilde{l}_{k+1} + \tilde{l}_1 f_{k+1}.$$

Analogously we obtain for $k = 1$:

$$-\delta f_1 = -\delta(j + c) = -\delta j + j^*\tilde{l}_2 + \delta j + \tilde{l}_1 f_2 = j^*\tilde{l}_2 + \tilde{l}_1 f_2 = f_1^*\tilde{l}_2 + \tilde{l}_1 f_2.$$

Thus, by the remarks preceeding Lemma 4.11 the family of maps $\{f_1, .., f_n\}$ forms an L_∞-morphism and the claim of the theorem is proven. □

Let us now apply our results to the symplectic case to retrieve the classical results:

Example 4.14. Let (M, ω) be a symplectic manifold and $\zeta : \mathfrak{g} \to \mathfrak{X}(M, \omega)$ an infinitesimal symplectic (right) action. Then ζ is Hamiltonian if and only if $\gamma\zeta = 0$. Assume now that ζ is Hamiltonian with a linear lift $j : \mathfrak{g} \to C^\infty(M)$. Then ζ is strongly Hamiltonian if and only if $[[j^*\{\cdot,\cdot\}]_{dR}]_{CE} = [j^*\{\cdot,\cdot\}]_{CE} = 0 \in H^2(\mathfrak{g}) \otimes H^0_{dR}(M)$.

Let us return to the examples from Section 3.1 in the light of n-plectic group actions:

Example 4.15. Let us continue Example 3.11. Recall from Theorem 21.1. in [4] that for a semi-simple Lie algebra $\mathfrak{g}$ the following holds:

- $H^1(\mathfrak{g}) = 0$ (equivalently $\mathfrak{g} = [\mathfrak{g}, \mathfrak{g}]$).
- $H^2(\mathfrak{g}) = 0$.
- $H^3(\mathfrak{g}) \neq 0$ and $[\omega_e]_{CE} = [\langle[\cdot,\cdot],\cdot\rangle]_{CE} \neq 0 \in H^3(\mathfrak{g})$, where $\langle\cdot,\cdot\rangle$ again denotes the (nondegenerate, symmetric, $\mathfrak{g}$-invariant) Killing form of $\mathfrak{g}$.

Here G acts on itself from the right by $(g, x) \mapsto x \cdot g$. The corresponding infinitesimal action ζ extends an $x \in \mathfrak{g}$ to a left-invariant vector field $\zeta(x)$. As $\mathfrak{g} = [\mathfrak{g}, \mathfrak{g}]$ the action is Hamiltonian and therefore it exists a linear lift j of ζ. Since ω is bi-invariant we obtain:

$$(j^* l_3)(X, Y, Z) = -(-1)^{\frac{2\cdot 3}{2}} \omega(\zeta(X), \zeta(Y), \zeta(Z)) = \omega_e(X, Y, Z).$$

So $[j^* l_3]_{CE} = [\omega_e]_{CE} \neq 0 \in H^3(\mathfrak{g}) = H^3(\mathfrak{g}) \otimes H^0_{dR}(G)$ and thus ζ can not be lifted to a strongly Hamiltonian action.

Before going to exterior powers of cotangent bundles we need the following result from [5]:

Lemma 4.16. Let $\zeta : \mathfrak{g} \to \mathfrak{X}(M, \omega)$ be an infinitesimal pre-n-plectic action. If ω has a $\mathfrak{g}$-invariant potential η then the action is strongly Hamiltonian and the L_∞-morphism takes the form $f_1(X) = (\zeta(X), \iota_{\zeta(X)}\eta)$ and for $k \geq 2$, $f_k(X_1, ..., X_k) = (-1)^k(-1)^{\frac{k(k+1)}{2}} \iota_{\zeta(X_k)} ... \iota_{\zeta(X_1)}\eta$.

Proof. To prove the statement we show, that the f_k defined in the lemma form an L_∞-morphism. First we show, that the equation $-\delta f_{k-1} = \tilde{l}_1 f_k + f_1^* \tilde{l}_k$ holds for $k = 2$:

$$\begin{aligned} -\delta f_1(X, Y) &= f_1([X, Y]) = \iota_{\zeta([X,Y])}\eta = \mathcal{L}_{\zeta(X)}\iota_{\zeta(Y)}\eta - \iota_{\zeta(Y)}\mathcal{L}_{\zeta(X)}\eta \\ &= \mathcal{L}_{\zeta(X)}\iota_{\zeta(Y)}\eta = (d\iota_{\zeta(X)} + \iota_{\zeta(X)}d)\iota_{\zeta(Y)}\eta = d\iota_{\zeta(X)}\iota_{\zeta(Y)}\eta - \iota_{\zeta(X)}\iota_{\zeta(Y)}d\eta \\ &= \tilde{l}_1 f_2(X, Y) + f_1^* \tilde{l}_2(X, Y) \end{aligned}$$

Using $\pi_{\mathfrak{X}} f_1 = \zeta$ we immediately obtain, that the equality also holds on the vector field component.

Next, using Equation 17 for η instead of ω, we show that $-\delta f_{k-2} = \tilde{l}_1 f_{k-1} + f_1^* \tilde{l}_{k-1}$ implies $-\delta f_{k-1} = \tilde{l}_1 f_k + f_1^* \tilde{l}_k$. In fact, by direct computation:

$$- \delta f_{k-1}(X_1, ..., X_k) = -(-1)^{(k-1)}(-1)^{(k-1)k/2} \sum_{1 \leq i < j \leq k} (-1)^{i+j} \iota_{\zeta(X_k)} ... \hat{\iota}_{\zeta(X_j)} ... \hat{\iota}_{\zeta(X_i)} ... \iota_{\zeta(X_1)} \iota_{\zeta([X_i, X_j])}\eta$$

$$= -(-1)^{(k-1)}(-1)^{(k-1)k/2} \sum_{1 \leq i < j \leq k} (-1)^{i+j} \iota_{\zeta(X_k)} ... \hat{\iota}_{\zeta(X_j)} ... \hat{\iota}_{\zeta(X_i)} ... \iota_{\zeta(X_1)} \iota_{[\zeta(X_i), \zeta(X_j)]}\eta$$

$$\begin{aligned} = -(-1)^{(k-1)}(-1)^{(k-1)k/2} \Bigg(&\sum_{1 \leq i < j < k} (-1)^{i+j} \iota_{\zeta(X_k)} ... \hat{\iota}_{\zeta(X_j)} ... \hat{\iota}_{\zeta(X_i)} ... \iota_{\zeta(X_1)} \iota_{[\zeta(X_i), \zeta(X_j)]}\eta \\ &+ \sum_{1 \leq i \leq k} (-1)^{i+k} \iota_{\zeta(X_{k-1})} ... \hat{\iota}_{\zeta(X_i)} ... \iota_{\zeta(X_1)} \iota_{[\zeta(X_i), \zeta(X_k)]}\eta \Bigg) \end{aligned}$$

$$\begin{aligned} = -(-1)^{k-1}(-1)^{(k-1)k/2} \Bigg(&(-1)^{k-2}(-1)^{(k-2)(k-1)/2} \iota_{\zeta(X_k)} \delta f_{k-2}(X_1, ..., X_{k-1}) \\ &+ \sum_{1 \leq i < k} (-1)^{i+k} \iota_{\zeta(X_{k-1})} ... \hat{\iota}_{\zeta(X_i)} ... \iota_{\zeta(X_1)} \iota_{[\zeta(X_i), \zeta(X_k)]}\eta \Bigg) \end{aligned}$$

$$= (-1)^{k+1}\iota_{\zeta(X_k)}\delta f_{k-2}(X_1, ..., X_{k-1}) + (-1)^{(k-1)k/2}\sum_{1\le i<k}(-1)^i\iota_{\zeta(X_{k-1})}...\hat{\iota}_{\zeta(X_i)}...\iota_{\zeta(X_1)}\iota_{[\zeta(X_i),\zeta(X_k)]}\eta$$

$$= (-1)^k\iota_{\zeta(X_k)}\left(\tilde{l}_1 f_{k-1}(X_1, ..., X_{k-1}) + f_1^*\tilde{l}_{k-1}(X_1, ..., X_{k-1})\right) + (-1)^{(k-1)k/2}\sum_{1\le i<k}(-1)^i\iota_{\zeta(X_{k-1})}...\hat{\iota}_{\zeta(X_i)}...\iota_{\zeta(X_1)}\iota_{[\zeta(X_i),\zeta(X_k)]}\eta$$

$$= (-1)^k\left((-1)^{k-1}(-1)^{(k-1)k/2}\iota_{\zeta(X_k)}\tilde{l}_1\iota_{\zeta(X_{k-1})}...\iota_{\zeta(X_1)}\eta - (-1)^{(k-1)k/2}\iota_{\zeta(X_k)}\iota_{\zeta(X_{k-1})}\iota_{\zeta(X_1)}\omega\right) + (-1)^{(k-1)k/2}\sum_{1\le i<k}(-1)^i\iota_{\zeta(X_{k-1})}...\hat{\iota}_{\zeta(X_i)}...\iota_{\zeta(X_1)}\iota_{[\zeta(X_i),\zeta(X_k)]}\eta$$

$$= \left(-(-1)^{(k-1)k/2}\iota_{\zeta(X_k)}\tilde{l}_1\iota_{\zeta(X_{k-1})}...\iota_{\zeta(X_1)}\eta - (-1)^{k(k+1)/2}\iota_{\zeta(X_k)}\iota_{\zeta(X_{k-1})}\iota_{\zeta(X_1)}\omega\right) + (-1)^{(k-1)k/2}\sum_{1\le i<k}(-1)^i\iota_{\zeta(X_{k-1})}...\hat{\iota}_{\zeta(X_i)}...\iota_{\zeta(X_1)}\iota_{[\zeta(X_i),\zeta(X_k)]}\eta$$

$$= \left(-(-1)^{(k-1)k/2}\iota_{\zeta(X_k)}\tilde{l}_1\iota_{\zeta(X_{k-1})}...\iota_{\zeta(X_1)}\eta + f_1^*\tilde{l}_k(X_1, ..., X_k)\right) + (-1)^{(k-1)k/2}\sum_{1\le i<k}(-1)^i\iota_{\zeta(X_{k-1})}...\hat{\iota}_{\zeta(X_i)}...\iota_{\zeta(X_1)}\iota_{[\zeta(X_i),\zeta(X_k)]}\eta$$

$$= f_1^*\tilde{l}_k(X_1, ..., X_k) + (-1)^{(k-1)k/2}\Bigg(- \iota_{\zeta(X_k)}\tilde{l}_1\iota_{\zeta(X_{k-1})}...\iota_{\zeta(X_1)}\eta + \sum_{1\le i<k}(-1)^i\iota_{\zeta(X_{k-1})}...\hat{\iota}_{\zeta(X_i)}...\iota_{\zeta(X_1)}\iota_{[\zeta(X_i),\zeta(X_k)]}\eta\Bigg)$$

$$= f_1^*\tilde{l}_k(X_1, ..., X_k) + (-1)^{(k-1)k/2}\left(\tilde{l}_1\iota_{\zeta(X_k)}\iota_{\zeta(X_{k-1})}...\iota_{\zeta(X_1)}\eta\right)$$

$$= f_1^*\tilde{l}_k(X_1, ..., X_k) + (-1)^{(k-1)k/2}\tilde{l}_1\iota_{\zeta(X_k)}\iota_{\zeta(X_{k-1})}...\iota_{\zeta(X_1)}\eta$$

$$= f_1^*\tilde{l}_k(X_1, ..., X_k) + (-1)^{(k-1)k/2}(-1)^k(-1)^{k(k+1)/2}\tilde{l}_1 f_k(X_1, ..., X_k)$$

$$= f_1^*\tilde{l}_k(X_1, ..., X_k) + \tilde{l}_1 f_k(X_1, ..., X_k).$$

Thus $-\delta f_{k-1} = \tilde{l}_1 f_k + f_1^*\tilde{l}_k$ holds for $k \in \{2, ..., n+1\}$, which proves the claim. □

Example 4.17. Continuation of Example 3.10.
Let G be a Lie group and $\vartheta^Q : M \times G \to Q$ a right action. For each g the map $\vartheta_g^Q : Q \to Q$ is a diffeomorphism. Then $T\vartheta_g^Q : TQ \to TQ$ is a fiberwise linear diffeomorphism, which makes the following diagram commute:

$$\begin{array}{ccc} TQ & \xrightarrow{T\vartheta_g^Q} & TQ \\ \downarrow & & \downarrow \\ Q & \xrightarrow{\vartheta_g^Q} & Q \end{array}$$

With the map $T\vartheta_g^Q$ at hand we construct a diffeomorphism $\Lambda^n T^*(\vartheta_g^Q) : \Lambda^n T^*Q \to \Lambda^n T^*Q$. Let η be an element of $\Lambda^n T^*Q$ with $\pi(\eta) = p \in Q$ and $v_1, ..., v_n \in T_{\vartheta_g^Q p}Q$.

$$(\Lambda^n T^*(\vartheta_g^Q))(\eta)(v_1, ..., v_n) = \eta((T_p\vartheta_g^Q)^{-1}v_1, ..., (T_p\vartheta_g^Q)^{-1}v_n),$$

where $(T_p\vartheta_g^Q)^{-1} : T_{\vartheta_g^Q p}Q \to T_pQ$ is the inverse of the linear map $T_p\vartheta_g^Q : T_pQ \to T_{\vartheta_g^Q p}Q$. Then $\vartheta_g^M := \Lambda^n T^*(\vartheta_g^Q)$ defines a right action which makes the following diagram commute:

$$\begin{array}{ccc} \Lambda^n T^*Q & \xrightarrow{\vartheta_g^M} & \Lambda^n T^*Q \\ \downarrow \pi & & \downarrow \pi \\ Q & \xrightarrow{\vartheta_g^Q} & Q \end{array}$$

Thus we have a right action ϑ^M of G on $M = \Lambda^n T^*Q$. To see that the action is n-plectic and even strongly Hamiltonian with respect to the canonical n-plectic structure ω it suffices to show that the n-form θ is G-invariant. Regard $\eta \in \Lambda^n T^*Q$ and $\xi_1, ..., \xi_n \in T_\eta(\Lambda^n T^*Q)$,

$$\begin{aligned}
((\vartheta_g^M)^*\theta)_\eta(\xi_1, ..., \xi_n) &= \theta_{\vartheta_g^M \eta}((T\vartheta_g^M)\xi_1, ..., (T\vartheta_g^M)\xi_n) \\
&= \vartheta_g^M(\eta)((T\pi)(T\vartheta_g^M)\xi_1, ..., (T\pi)(T\vartheta_g^M)\xi_n) \\
&= \vartheta_g^M(\eta)((T(\pi \circ \vartheta_g^M))\xi_1, ..., (T(\pi \circ \vartheta_g^M))\xi_n) \\
&= \vartheta_g^M(\eta)((T(\vartheta_g^Q \circ \pi))\xi_1, ..., (T(\vartheta_g^Q \circ \pi))\xi_n) \\
&= \eta((T\vartheta_g^Q)^{-1}(T(\vartheta_g^Q \circ \pi))\xi_1, ..., (T\vartheta_g^Q)^{-1}(T(\vartheta_g^Q \circ \pi))\xi_n) \\
&= \eta((T\pi)\xi_1, ..., (T\pi)\xi_n) \\
&= \theta_\eta(\xi_1, ..., \xi_n).
\end{aligned}$$

Thus $(\vartheta_g^M)^*\theta = \theta$ and thus ω is G-invariant with an invariant potential. Lemma 4.16 now implies that the action is strongly Hamiltonian with homotopy co-moment map defined via the G-invariant potential $\eta = -\theta$ of ω.

Remark 4.18. As hinted at in [5], the components of a strong homotopy co-moment map can be interpreted as certain "momentum maps" previously discussed in the literature. We expand and generalize their agrument here.

- The first component f_1 of a strong co-moment map on an n-plectic manifold (M, ω) corresponds to a covariant multimomentum map in the sense of [3]. There an adapted version of Diagram (18), with exact forms divided out, is used:

$$0 \longrightarrow H^{n-1}_{dR}(M) \longrightarrow \frac{\Omega^{n-1}_{Ham}(M,\omega)}{\Omega^{n-1}_{ex}(M)} \longrightarrow \mathfrak{X}_{Ham}(M,\omega) \longrightarrow 0 \quad .$$

As the term obstructing the Jacobi identity for l_2 is here exact, $\frac{\Omega^{n-1}_{Ham}(M,\omega)}{\Omega^{n-1}_{ex}(M)}$ carries a Lie structure induced by l_2. Together with the trivial bracket on $H^{n-1}_{dR}(M)$ this bracket turns the sequence into an exact sequence of Lie algebras. A *covariant multimomentum map* in the sense of [3] for an infinitesimal (weakly) Hamiltonian action $\zeta : \mathfrak{g} \to \mathfrak{X}_{Ham}(M,\omega) \subset \mathfrak{X}(M,\omega)$ is the dual of a linear lift $f : \mathfrak{g} \to \frac{\Omega^{n-1}_{Ham}(M,\omega)}{\Omega^{n-1}_{ex}(M)}$ of ζ. The class $[c] \in H^2(\mathfrak{g}, H^{m-1}_{dR}(M))$ in Section 4.2 of [3] obstructing f to be a Lie algebra morphism is exactly our class h_1.

- The last component f_n of a strong co-moment map is equivalent to a multi-moment map in the sense of [8].

  A *multi-moment map* is a map $v : M \to P^*_{\mathfrak{g}}$, where $P_{\mathfrak{g}} = P_{\mathfrak{g},n} \subset \Lambda^n\mathfrak{g}$ is the kernel of $D : \Lambda^n\mathfrak{g} \to \Lambda^{n-1}\mathfrak{g}$, satisfying certain conditions. It can be equivalently understood as a map $\overline{v} : P_{\mathfrak{g}} \to C^\infty(M)$ fulfilling

  (i) $d(\overline{v}(p)) = \iota_{\zeta(p)}\omega$ for all $p \in P_{\mathfrak{g}}$

  (ii) $\overline{v}(ad_X(p)) = \mathcal{L}_{\zeta(X)}(\overline{v}(p))$ for all $X \in \mathfrak{g}$ and $p \in P_{\mathfrak{g}}$.

  Here we define for $X, X^\alpha_j \in \mathfrak{g}$ and $p = \sum_\alpha X^\alpha_1 \wedge ... \wedge X^\alpha_n \in \Lambda^n\mathfrak{g}$,

$$\iota_{\zeta(p)}\omega := \sum_\alpha \iota_{\zeta(X^\alpha_n)}...\iota_{\zeta(X^\alpha_1)}\omega \quad \text{and} \quad ad_X(p) := (\Lambda^n(ad_X))(p) \ .$$

  Given a strong co-moment $\{f_1, ..., f_n\}$, we put $\overline{v} := -(-1)^{n(n+1)/2} f_n|_{P_{\mathfrak{g}}}$ and claim that it fulfills the two preceding conditions. Let $p = \sum_\alpha X^\alpha_1 \wedge ... \wedge X^\alpha_n$ be an element of $P_{\mathfrak{g}}$, then

$$\begin{aligned} d(\overline{v}(p)) &= (-1)^{n(n+1)/2}\tilde{l}_1 f_n(p) = -(-1)^{n(n+1)/2}((\delta f_{n-1})(p) + f_1^*\tilde{l}_n(p)) \\ &= -(-1)^{n(n+1)/2}(-f_{n-1}(Dp) + f_1^*\tilde{l}_n(p)) = -(-1)^{n(n+1)/2} f_1^*\tilde{l}_n(p) = \iota_{\zeta(p)}\omega, \end{aligned}$$

  since p is annihilated by $D = -\delta^*$.

  Condition (ii) is equivalent to

$$\mathcal{L}_{\zeta(X)}(f_n(\sum_\alpha X^\alpha_1 \wedge ... \wedge X^\alpha_n)) = f_n(\sum_{1\le j\le n}\sum_\alpha X^\alpha_1 \wedge ... \wedge [X, X^\alpha_j] \wedge ... \wedge X^\alpha_n),$$

which we show to hold in the sequel. Let us first take a look at the left hand side:

$$\mathcal{L}_{\zeta(X)}(f_n(\sum_\alpha X_1^\alpha \wedge ... \wedge X_n^\alpha)) = (\iota_{\zeta(X)}d + d\iota_{\zeta(X)})f_n(\sum_\alpha X_1^\alpha \wedge ... \wedge X_n^\alpha)$$

$$= \iota_{\zeta(X)}d\left(f_n(\sum_\alpha X_1^\alpha \wedge ... \wedge X_n^\alpha)\right) = \iota_{\zeta(X)}\left((-\delta f_{n-1} - f_1^*\tilde{l}_n)(\sum_\alpha X_1^\alpha \wedge ... \wedge X_n^\alpha)\right)$$

$$= \iota_{\zeta(X)}\left(-f_1^*\tilde{l}_n(\sum_\alpha X_1^\alpha \wedge ... \wedge X_n^\alpha)\right) = \iota_{\zeta(X)}(-1)^{n(n+1)/2}\sum_\alpha \iota_{\zeta(X_n^\alpha)}...\iota_{\zeta(X_1^\alpha)}\omega$$

$$= (-1)^{n(n+1)/2}\sum_\alpha \iota_{\zeta(X)}\iota_{\zeta(X_n^\alpha)}...\iota_{\zeta(X_1^\alpha)}\omega = -(-1)^{n(n+1)/2}(-1)^{(n+1)(n+2)/2}f_1^*\tilde{l}_{n+1}(\sum_\alpha X_1^\alpha \wedge ... \wedge X_n^\alpha \wedge X)$$

$$= (-1)^n f_1^*\tilde{l}_{n+1}(\sum_\alpha X_1^\alpha \wedge ... \wedge X_n^\alpha \wedge X).$$

For the right hand side we have (setting $X_0^\alpha := X$ for all α and using again that $0 = D(p)$ $= \sum_\alpha \sum_{1\le i<j\le n}(-1)^{i+j}[X_i^\alpha, X_j^\alpha] \wedge X_1^\alpha ... \wedge \hat{X}_i^\alpha ... \hat{X}_j^\alpha ... \wedge X_n^\alpha$):

$$f_n\left(\sum_\alpha \sum_{1\le j\le n} X_1^\alpha \wedge ... \wedge [X, X_j^\alpha] \wedge ... \wedge X_n^\alpha\right) = f_n\left(\sum_\alpha \sum_{1\le j\le n} (-1)^{j-1}[X, X_j^\alpha] \wedge X_1^\alpha \wedge ... \wedge \hat{X}_j^\alpha \wedge ... \wedge X_n^\alpha\right)$$

$$= f_n\Bigg(-\sum_\alpha \sum_{1\le j\le n} (-1)^j[X, X_j^\alpha] \wedge X_1^\alpha \wedge ... \wedge \hat{X}_j^\alpha \wedge ... \wedge X_n^\alpha$$

$$- X \wedge \sum_\alpha \sum_{1\le i<j\le n} (-1)^{i+j}[X_i^\alpha, X_j^\alpha] \wedge X_1^\alpha ... \wedge \hat{X}_i^\alpha ... \hat{X}_j^\alpha ... \wedge X_n^\alpha\Bigg)$$

$$= f_n\left(-\sum_\alpha \sum_{0\le i<j\le n} (-1)^{i+j}[X_i^\alpha, X_j^\alpha] \wedge X_0^\alpha ... \wedge \hat{X}_i^\alpha ... \hat{X}_j^\alpha ... \wedge X_n\right)$$

$$= (-\delta f_n)\left(X \wedge \sum_\alpha X_1^\alpha \wedge ... \wedge X_n^\alpha\right) = (-1)^n(-\delta f_n)\left(\sum_\alpha X_1^\alpha \wedge ... \wedge X_n^\alpha \wedge X\right).$$

Hence the equivariance of $\bar{\nu}$ follows from the fact that $-\delta f_n = f_1^*\tilde{l}_{n+1}$ (upon recalling that $f_{n+1} = 0$). Thus $\bar{\nu}$ defines a multi-moment map in the sense of [8].

A Universal coalgebra properties

In the course of reformulating the two definitions of an L_∞-algebra we often needed statements about a map being uniquely extendable or uniquely determined by its values on the cogenerators. This relies on the following theorem and corollary, which are proven in Appendix B, Part 4 of [10]:

Theorem A.1. Let C be a connected differential graded coaugmented co-commutative coalgebra. Then $\theta \mapsto \pi_1 \circ \theta$ is a bijection from the set of coalgebra maps $\theta : C \to S^\bullet V$ to differential graded linear maps $u : C \to V$ which satisfy $u(1) = 0$, where $1 := i_0(1)$ and i_0 is the coaugmentation map of C.

Corollary A.2. Let M be a differential graded comodule under $S^\bullet V$. Then there is a one-to-one correspondance between degree r coderivations $\delta : M \to S^\bullet V$ and degree r graded linear maps $v : M \to V$ given by $\delta \mapsto \pi_1 \circ \delta$.

Being coaugmented means that, apart from having a counit the coalgebra also has a map $i_0 : \mathbb{K} \to C$ such that the following diagrams commute:

compatibility with diagonal:

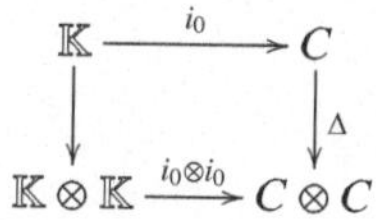

compatibility with counit:

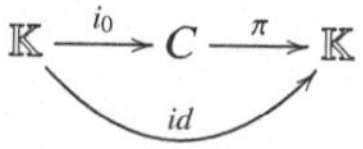

A coaugmented coalgebra is called connected if $C = \bigcup F_k C$, where F_k is defined recursively via the formulas: $F_0 C = i_0(\mathbb{K})$, $F_k C = \{x \in C \mid (\Delta x - 1 \otimes x - x \otimes 1) \in F_{k-1} C \otimes F_{k-1} C\}$. One can check, that $S^\bullet V$ itself is coaugmented and connected. See for example [5] for a detailed discussion.

References

[1] Michael P. Allocca. *L-infinity Algebra Representation Theory*. ProQuest LLC, Ann Arbor, MI, 2010. Thesis (Ph.D.)–North Carolina State University.

[2] John C. Baez and Alissa S. Crans. Higher-dimensional algebra. VI. Lie 2-algebras. *Theory Appl. Categ.*, 12:492–538, 2004.

[3] José F. Cariñena, Mike Crampin, and Luis A. Ibort. On the multisymplectic formalism for first order field theories. *Differential Geom. Appl.*, 1(4):345–374, 1991.

[4] Claude Chevalley and Samuel Eilenberg. Cohomology theory of Lie groups and Lie algebras. *Trans. Amer. Math. Soc.*, 63:85–124, 1948.

[5] Yael Fregier, Christopher L. Rogers, and Marco Zambon. Homotopy moment maps. `arXiv preprint arXiv:1304.2051`, 2013.

[6] Tom Lada and Martin Markl. Strongly homotopy Lie algebras. *Comm. Algebra*, 23(6):2147–2161, 1995.

[7] Tom Lada and Jim Stasheff. Introduction to SH Lie algebras for physicists. *Internat. J. Theoret. Phys.*, 32(7):1087–1103, 1993.

[8] Thomas B. Madsen and Andrew Swann. Closed forms and multi-moment maps. *Geom. Dedicata*, 165:25–52, 2013.

[9] Peter W. Michor. *Topics in differential geometry*, volume 93 of *Graduate Studies in Mathematics*. American Mathematical Society, Providence, RI, 2008.

[10] Daniel Quillen. Rational homotopy theory. *Ann. of Math. (2)*, 90:205–295, 1969.

[11] Christopher L. Rogers. *Higher Symplectic Geometry*. ProQuest LLC, Ann Arbor, MI, 2011. Thesis (Ph.D.)–University of California, Riverside.

[12] Urs Schreiber. L-infinity-algebra. `http://ncatlab.org/nlab/show/L-infinity-algebra`, 2014.

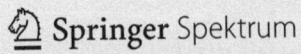